Food Flavorings, Colorings and Preservatives: A Chef's Guide

Food Flavorings, Colorings and Preservatives: A Chef's Guide

Manuel Newman

STATES
ACADEMIC PRESS
www.statesacademicpress.com

Food Flavorings, Colorings and Preservatives: A Chef's Guide
Manuel Newman
ISBN: 978-1-63989-801-5 (Hardback)

Published by States Academic Press,
109 South 5th Street,
Brooklyn, NY 11249, USA

Visit us on the World Wide Web at:
www.statesacademicpress.com

Cataloging-in-Publication Data

Food flavorings, colorings and preservatives : a chef's guide / Manuel Newman.
p. cm.
Includes bibliographical references and index.
ISBN 978-1-63989-801-5
1. Cooks--Vocational guidance. 2. Flavoring essences--Vocational guidance.
3. Coloring matter in food--Vocational guidance. 4. Food preservatives--Vocational guidance.
I. Newman, Manuel.
TX652.4 .F66 2023
641.502--dc23

Table of Contents

Preface

Flavor refers to the sensory impression of food, which is determined by the sense of smell and taste. Smell is the main determinant of the flavor of a food item. The basic tastes of bitter, sweet, sour and salty are recognized universally. Some cultures also recognize pungency and oleogustus. The flavor of food can be altered or enhanced with both natural and artificial flavorants. Food colors or color additives are pigments, dyes or substances that impart color when added to a food or drink. They can be liquids, pastes, powders or gels. People associate colors with particular flavors and hence, food colors can influence the perceived flavor of the food. Food preservatives are added to food to prevent spoilage and visual deterioration by inhibiting the growth of microorganisms. Sugaring, jellying, canning, pickling, and fermentation are some of the traditional techniques of food preservation. This textbook is compiled in such a way that it will provide in-depth knowledge about the use and application of food flavorings, food preservatives and food colors. Most of the topics covered herein address the modern trends in food preparation, stylization and preservation. It will prove to be an important reference text to students, chefs and anyone else with a passion for cooking.

A short introduction to every chapter is written below to provide an overview of the content of the book:

Chapter 1 - Food flavor refers to the sensory perception of food and other substances determined by its taste and smell. It can be altered with artificial or natural flavorants. The principal kinds of flavoring are natural flavoring substances, nature-identical flavoring substances and artificial flavoring substances. This chapter provides an overview of food flavoring and the ways of enhancing a food's taste and flavor; **Chapter 2 -** Flavorings are used for altering the flavors of natural foods such as vegetables and meats. These are also used for creating flavors for items such as candies and snacks, which do not have desired flavors. This chapter explores the different ways to flavor cake, rice, chicken, butter, turkey, etc. for attaining a balance of flavors in cooking; **Chapter 3 -** Seasoning is a technique in food preparation in which salt, herbs or spices are added to food to enhance its flavor. Other seasonings like sea salt, black pepper, basil and seasoning sauce can also be used. This chapter examines the various processes of improving flavor through an understanding of different seasonings and seasoning techniques; **Chapter 4 -** Food coloring is any substance that imparts color to a food or drink. They can be in the form of liquids, gels, pastes and powders. Color additives are added to food to improve its visual appeal, enhance natural colors or offset color loss. The topics elucidated in this chapter on food coloring cover the various aspects of making natural food coloring, black food coloring, brown food coloring, vegetable dyes for Easter Eggs, etc.; **Chapter 5 -** Food preservation is aimed at preventing the growth of microorganisms, slowing down the oxidation of fats causing rancidity and inhibiting the visual deterioration of food. Several methods of preservation of food exist such as drying, dehydrating, smoking, freezing and vacuum packing. This chapter discusses these preservation methods and their effective uses; **Chapter 6 -** Food presentation is the art of decorating, processing, arranging and modifying food for enhancing its aesthetic appeal. Foods may be decorated with icing, drizzled with sauces, topped with edible ornamentations, sprinkled with toppings, powders and seeds, or garnished. The various techniques of food presentation along with the aspects of plating of food, using Nasturtiums in food, etc. have been elaborated in this chapter.

Finally, I would like to thank my fellow scholars who gave constructive feedback and my family members who supported me at every step.

Manuel Newman

Chapter 1

Introduction to Food Flavoring

Food flavor refers to the sensory perception of food and other substances determined by its taste and smell. It can be altered with artificial or natural flavorants. The principal kinds of flavoring are natural flavoring substances, nature-identical flavoring substances and artificial flavoring substances. This chapter provides an overview of food flavoring and the ways of enhancing a food's taste and flavor.

Flavorings are intense preparations which are added to foods in order to impart taste and/or smell. These food flavors are used in small amounts and are not intended to be consumed alone. There are certain natural food flavors which are derived from herbs, spices and substances having an exclusively sweet, sour or salt taste. These natural food flavors are not included in the definition of flavorings for regulatory purposes.

Types of Food Flavoring

Flavorings are used as food additives for altering and/or enhancing the flavors of natural food products. Sometimes, food flavorings are also used to create flavor for food products that do not have desired flavors such as candies and other snacks. There are three major types of food flavorings that are used in foods. These types of flavorings are mostly used as criteria for food regulatory purposes in European Union and Australia. In North America, the classification of flavorings is done as - Natural flavorings and Synthetic flavorings. The Synthetic flavorings there include both artificial and nature-identical flavorings.

Natural Flavoring Substances

Flavoring substances that are obtained from plant or animal raw materials, by physical, microbiological or enzymatic processes are classified as natural flavoring substances. These natural flavorings can be either used in their natural form or processed form for consumption by human beings. However, they cannot contain any nature-identical or artificial flavoring substances.

Nature-identical Flavoring Substances

Nature-identical substances are the flavoring substances that are obtained by synthesis or are isolated through chemical processes, which are chemically identical to flavoring substances naturally present in products intended for consumption by human beings. These flavorings cannot contain any artificial flavoring substances.

Artificial Flavoring Substances

Flavoring substances that are not identified in a natural product intended for consumption by human being- whether or not the product is processed- are artificial flavoring substances. These food flavorings are typically produced by fractional distillation and additional chemical manipulation naturally sourced chemicals or from crude oil or coal tar.

Chemicals associated with particular flavors

Allylpyrazine	Flavor
Color	Roasted nut
Methoxypyrazines	Earthy vegetables
2-Isobutyl-3 Methoxypyrazine	Green pepper
Acetyl-L-Pyrazines	Popcorn
Aldehydes	Fruity, green
Alcohols	Bitter, medicinal
Esters	Fruity
Ketones	Butter, caramel
Pyrazines	Brown, burnt, caramel,
Phenolics	Medicinal, smokey
Terpenoids	Citrus, piney

Smoke Flavoring Substances

Although it is not one of the basic flavorings, smoke flavoring has come up as a significant food flavoring substance in the world of food additives. Smoke flavoring is a natural flavoring concentrate obtained by subjecting untreated and uncontaminated hardwood, including sawdust and woody plants, to one or more of the se listed processes for obtaining fractions which have the desired flavor potential.

- Controlled burning.
- Dry distillation at appropriate temperatures.
- Treatment with superheated steam.

Process of Flavor Creation

The oldest and simplest method for flavor creation is the artistic approach. The second approach is to combine art with scientific know-how. The third approach is to follow nature's footsteps and develop flavors biosynthetically.

Artistic Approach

In this approach, the artistic powers of the flavor chemist are at work. These several components comprise. The sensory characteristics of flavor components like synthetic organic chemicals, natural essential oils, botanical extracts, juices, spices, nuts, herbs, and concentrates of animal origin, leave various impressions. Each flavor chemist will have the characteristics of these components and their corresponding food-nuance associations stored in memory, dependent on his or her personal ability and training.

This approach can be called an iterative method, or a trial by error approach. The flavorist produces a trial formulation, evaluates the trial by taste, and then makes revisions based upon their evaluation. This is the traditional type of methodology and has been used by flavorists for many years.

Scientific Approach

The artistic approach is used along with the scientific knowledge. The utilization of chromatography, along with various other analytical techniques, has the flavor chemist gain a better understanding of nature's process in producing flavors. This knowledge is used in combination with the artistic skills of the flavor chemist to create unique flavors.

Biosynthetic Approach

In this approach, the biochemist attempts to duplicate nature's biogenetic pathways. Although science has not yet unlocked many of nature's secrets in developing its flavors, some of the known enzymatic and fermentation reactions are crudely exploited to produce building blocks. Enzymatically modified cheeses, fermented fruits and wine, and cooked and roasted foods are just a few examples. Due to the increased demand for natural flavors, research in this area is dramatically increasing.

The primary areas of biotechnology that are being utilized and explored for use in the flavor business are:

i. The product of natural flavor ingredients.

ii. The production of part or complete flavors.

iii. Increased yields of essential oils, oleoresin, and flavor components by the use of enzymes in the processing of natural materials.

iv. Plant cloning for maximum yields of secondary metabolites.

v. Specific chemical steps to produce economies in the production of expensive flavor chemicals or to support natural claims (e.g., production of natural benzaldehyde from other natural chemicals).

vi. Modified traditional fermentation systems to produce flavor enhanced bases for direct use or further concentration.

vii. Secondary metabolites from in vitro tissue culture of vegetable cells.

Table: Examples of biosynthetic production of chemicals

Sr. No.	Major product(s) produced in nature	Examples of occurrence	Microorganism applied	Major portion of substrate
1	Methyl ketones	Cheese	Penicillium Roqueforti	Fatty acids
2	Lactones	Peaches, coconut	Pityrosporum species	Lipids

3	Butyric acid	Butter	Clostridium Butyricum	Dextrose
4	Carveol, carvone, dihydrocarvone, perillyl alcohol	Spearmint and other essential oils	Pseudomonas species	Limonene
5	d-Verbenone, d-cis-Verbenol	-	Aspergillus Niger	α-Pinene
6	Cheese like flavor	Cheese	Streptococcus species Lactobacillus species	Reconstituted milk
7	Bread like flavor	Bread	Saccharomyces Cerevisiae	Sugar and milk

Thermally Produced Flavors

In this process, the chemist follows the natural route of forming the "secondary origin" flavors mentioned in table. Ingredients used in their preparation consist of one or more of the following:

1. A protein nitrogen source, for example, meat, poultry, dairy products, seafood, and their hydrolysis products.
2. A carbohydrate source, for example, vegetable, fruit, sugars, and their hydrolysis products.
3. A fat or fatty source.
4. Herbs, spices, water, selected vitamins, acid, emulsifiers, and nucleotides.
5. Flavoring preparations and flavor adjuncts to be added after processing.

The prepared mixture is then processed under controlled conditions. The temperature should not exceed 180°C for the specified time. The general concept of producing flavors thermally is to utilize foodstuffs or constituents of foodstuffs and apply processes comparable to traditional kitchen treatments.

Recent Flavor Creation Technologies

Flavor companies are using computer technology to assist the flavorist in their creative task. Many companies have introduced systems where a flavorist has access to a database of thousands of different flavor preparations. By doing selective searches through this database, the flavorist can retrieve a starting formulation which meets specific parameters (e.g., liquid, natural, kosher, and heat stable). From there, the flavorist can tailor a new formulation that meets a specific customer's needs. By the use of this type of technology, the time to develop a new flavor has been reduced significantly. There are also equipments developed, which allows flavorist to evaluate almost instantaneously the aroma of blends they make using computer technology even to the extent of including any aroma of a food product—for example, yogurt. Other equipment is used to accurately measure threshold levels either of individual components or of flavors in finished food products. By using this type of equipment, the flavorist, supported by sensory specialists, can evaluate the different release rates of a flavor in different bases or products and then can adjust formulations so that the same flavor perception can be achieved in different applications.

General Rules for Flavor Creation

In spite of the increasing number of technological processes and tools, artistry is still the most important factor in the process of flavor creation. Modern flavor chemists utilize all the scientific resources available to them to create the proper balance between art and science to achieve excellence. Attempts to use science only, such as using analytical tools to recreate flavors from data produced, have proven to be fruitless. Reasons for failure of such an approach can be summarized as follows:

i. Only a few of the hundreds of chemicals identified in foods have sensory significance.

ii. Availability and cost of the total chemicals identified will prove prohibitively costly.

iii. The methodology to obtain qualitative and quantitative identification of trace chemicals is not very accurate.

iv. Synthetic chemicals, utilized to duplicate nature, contain trace impurities that affect the final taste of the compound.

Therefore, technical resources are tools that should be employed to service the flavor chemist.

Important factors influencing flavor creation are:

a. Flavor Profile of the Target: The flavor target could be any flavor, aroma fruit such as a mango to roast beef. Drawing the preliminary flavor profile of this target is an important step in the process of the flavor development and constitutes the foundation upon which the flavor will be created.

b. Flavor Descriptive Language: To reduce communication gap between industry, customers and also amongst flavor chemists for description of a particular flavor, it is imperative to use a standard descriptive language among flavor chemists and food scientists.

c. Sensory Evaluation: Application of proper sensory techniques, and large panels will make the difference in the fine tuning process. Reviewing the flavor profile by utilizing focus groups will bring the right balance of art and science. Consumers should be the ultimate judges of the quality of a flavor.

d. Reproducibility: Since flavors are complex mixtures of various components, the flavor chemists' task is to ensure the compatibility of such mixtures and also it is important that any laboratory process should be reproducible in production.

Analytical Techniques for Determining Flavor Components

The process of determining the flavor components in a food can be divided into three general phases.

- Isolation of the volatile flavor components from the bulk of the nonvolatile matrix

Once the sample has been reduced to the proper physical form, the volatile flavor components are usually isolated by either distillation or extraction, or a combination of the two. Since most liquid samples are in an aqueous medium, a simple extraction with an organic solvent (either polar solvents such as methylene chloride or nonpolar solvents such as ether or pentane) in a separatory funnel are used.

There are also solvents that can be used in special cases, such as isopentane, which can extract the volatile components from a matrix high in ethanol or propylene glycol. soxhlet extraction apparatus is used for the extraction of solid samples, such as ground nuts or seeds. since this method collects the extracted volatiles in boiling solvent, heat-sensitive compounds may be lost or produce artifacts. Where significant heat-sensitive materials are present, an extraction in which solvent is continuously recirculated through the sample by means of a pump at ambient temperature is more suitable.

- Separation of the isolated components into individual or small groups of components

Column and thin-layer chromatography are generally used for rough separations into groups of components, while gas chromatography (GC) and high-performance liquid chromatography (HPLC) are of greater value in determining individual components.

- Identification of the flavor components

There are a wide variety of tools to identify unknown components. If the general class of the flavor components is known and compared to an accumulated library of standards, retention time data can be used for identification. Tables of retention indices such as the Kovats index have been published. Kovats indices provide the retention times of individual components compared with those of a homologous series of alkanes or esters. SNIFF analysis may also be used. In this method, an unknown sample is injected into the gas chromatograph and is separated into the various components. As the components are eluted, a portion of each is sent to the detector and a separate portion is also sent to a sniffing port where the flavorist evaluates the odor of the component as it is released from the gas chromatograph and identifies the component. Mass spectrometry (MS) is also a widely used tool to identify unknown components. The mass spectrometer is attached to a gas chromatograph which separates the unknown sample into its components. With the progress in computers GC-MS systems are nowadays available at reasonable costs.

Applications of Flavors

Flavors are applied to various products in the food, beverage, tobacco, pharmaceutical, and oral hygiene areas. Lately, some applications have extended to other segments such as the toy industry. Flavored products fall into two categories:

1. Flavor-dependent: These are foods and beverages that cannot exist without the application of flavors. Examples are hard-boiled candy, chewing gum, carbonated and nonjuice drinks, gelatin desserts, and powdered artificial beverages.
2. Flavor-independent: These are products that can be marketed without flavors (e.g. crackers, cereals, and nuts) or for which flavors are legally prohibited (e.g. milk, orange juice, and butter wherein flavor reinforcement is not permitted, unless a new identity is given to the food).

Direct and Indirect Flavoring

It is interesting to note that almost every food can be flavored provided that consumers are not misled or deceived. The food industry achieves this, as in the following examples:

1. New identity: A food such as milk, butter, or juice can be flavored provided that it is name

to convey to consumers its new identity and avoid consumer deception. Examples are blends of butter and margarine.

2. Indirect flavoring: Indirect flavorings are preparations that might be added by consumers to their home-cooked meals e.g. Salad dressings.

Flavor Selection

It is important for flavor users to understand the complexity of flavor selection. It will be helpful to begin by separating flavors from all other additives and placing them in a class by themselves. The main reason for the said purpose is Specificity. Flavors are very specific to certain applications. For example, the same orange flavor (same identification number and supplier) might taste excellent in some applications but be much less acceptable in others. Flavors in many instances are tailored to suit a complex food system as well as the processing conditions applied.

Flavor Forms

Flavor strength and potency vary considerably. Some flavors are diluted with solvents and carriers, whereas others are compounds of aromatic chemicals without a solvent. Various flavor forms are delineated in table.

Table: Forms of flavors available in market

Type	Forms	Solvents and carriers
Liquid	Water-, alcohol-, or oil-soluble	Alcohol, propylene glycol, triacetin, benzyl alcohol, glycerin, syrup, water, vegetable oil
Powder	Spray-dried, absorbates, or powder mixes	Gum acacia, starch hydrolysates, selective hydrocolloids, simple carbohydrates
Pastes and emulsions	Emulsion of the oil-in-water type	Same ingredients as for powder and liquid

Chapter 2 Flavoring Different Types of Food

Flavorings are used for altering the flavors of natural foods such as vegetables and meats. These are also used for creating flavors for items such as candies and snacks, which do not have desired flavors. This chapter explores the different ways to flavor cake, rice, chicken, butter, turkey, etc. for attaining a balance of flavors in cooking.

How to Flavor Chicken

Chicken can sometimes taste a little boring all by itself. There are plenty of ways to add some flavor, though. Use dry rubs and spice mixes to add flavor quickly and easily. Marinating and brining take a little extra time, but both result in flavorful, juicy chicken. You can also try a wide variety of glazes and sauces to coat your chicken with extra flavor.

Method 1. Using Dry Rubs and Spice Mixes

1. Buy pre-made spice mixes in a few different flavors. Spice mixes take all the guesswork out of seasoning by offering a perfectly balanced solution. There are tons of pre-made spice mixes available at any grocery store. You can achieve almost any flavor imaginable. Pick up a few different flavors and experiment to discover what you like best.

- Examples of spice mix flavors to try are peppercorn ranch, lemon pepper, Tuscan sunset, Southwest, and Cajun.
- Refer to the product directions for specific information about how much of the spice mix to use.

2. Experiment with your own combinations of spices. If you like to get creative in the kitchen, try whipping up your own custom blends. Find a typical spice blend recipe, then change up some of the ingredients to create a different flavor. For example, you could alter a Mexican spice mix by taking out the chili powder and cumin. Replace them with basil and garlic powder to create a spicy Italian blend.

- Keep your spice mixes in airtight containers and label them. Try labeling mixes with something fun like "Josh's Famous Cajun Spice Mix." Spice mixes will stay fresh for 1-2 years of stored correctly.
- The difference between spice mixes and dry rubs is the addition of brown sugar. Try experimenting with the kinds of sugar you use. For example, replace some of the brown sugar with white sugar or turbinado.

3. Whip up a basic dry rub for your chicken. Measure out your ingredients and add them to a large zip-lock bag. Seal the bag securely. Shake it for about 30 seconds until the ingredients combine evenly inside the bag. Rub the mixture on your chicken, wrap it up in plastic, and put it in your fridge overnight. When you're ready to cook, just unwrap the chicken and proceed with your dish.

- Once unwrapped, start making your dish immediately. The flavor of the rub will start to degrade if left sitting out unwrapped too long.

- You can leave it in the fridge for up to 72 hours if you want to experiment with stronger flavors.

Method 2. Brining and Marinating Chicken

1. Create a basic brine by combining salt and water. Brine is easy to make, and the result is great flavor and juicy chicken. The ratio of salt to water is 4 tablespoons of salt per 1 quart (4 cups) of water. Add both to a large bowl and stir until the salt dissolves. Submerge the chicken in the brine solution. Cover the bowl and refrigerate for one hour. Remove the bowl, drain the brine, and proceed with your dish as you normally would.

- Choose a bowl deep enough to fully submerge the chicken. Use the ratio to create more, if needed.
- For added flavor, try adding a handful of fresh herbs or a small amount of lemon zest to your brine.
- Brining for one hour is perfect if you're making a chicken dish for your family. If you're cooking a whole chicken, let it brine for about 12 hours.

2. Experiment with store-bought marinades. Just like spice mixes, your favorite grocery store will offer a wide variety of bottled marinades. Place your thawed chicken in a heavy-duty zip-lock bag, then pour in the marinade of your choice. The meat should be fully submerged in the marinade. Seal the bag tightly and put it in the fridge for 6 to 8 hours. When you're ready to cook, just drain the marinade and proceed with your dish.

- Marinades work especially well for chicken you plan to grill or bake.
- Examples of store-bought marinades you can try are pesto, orange sesame, and tropical teriyaki.

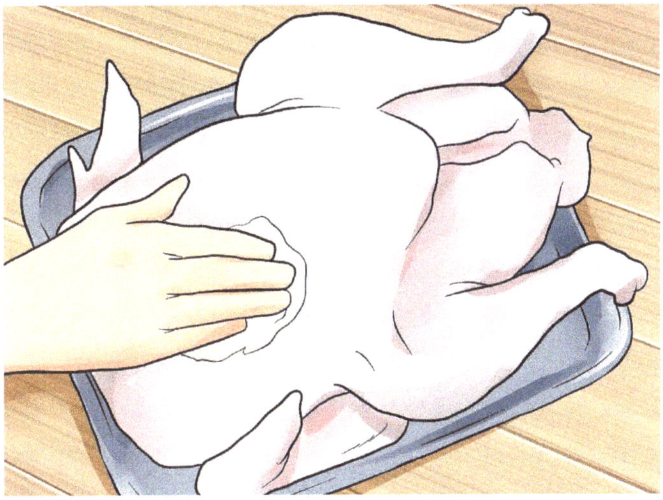

3. Soak your chicken in a spicy Korean marinade. Combine the onions, garlic, soy sauce, gochujang, brown sugar, honey, rice vinegar, and sesame oil in a large bowl. Place your chicken in a large zip-lock bag. Carefully pour the marinade mixture into the bag. Make sure the chicken is fully covered with the marinade. Seal the bag tightly. Place it in your fridge for 6 to 8 hours.

Method 3.Using Glazes and Sauces

1. Whip up your own flavorful sauce as a topping. Start with a handful of fresh herbs. You can use basil, oregano, or anything else you like. Chop the herbs up and add some citrus zest to a bowl. Mix in a small amount of olive oil, stir, and then top your chicken with it.

- You can use it while you're making the chicken, or even drizzle it over cooked chicken for an extra burst of flavor.
- You can also experiment with using premade pesto and chunky salsa.

2. Brush on a tasty glaze while you're cooking the chicken. Glazes can be as simple or as complex as you like. Try mixing Dijon mustard with maple syrup and brushing it on the chicken as it cooks. Another popular glaze combination is marmalade and soy sauce. You can even use something as simple as store-bought barbeque sauce.

- Reserve the remaining glaze and serve it with the chicken as a dipping sauce.

3. Put together a sweet and tangy BBQ Sauce. Add a little olive oil to a medium sized sauce pan over medium heat. Add the onions and cook for 5 minutes. Add the garlic and cook for another 1-2 minutes. Reduce the heat to low. Add the remaining ingredients to the pan. Stir and then simmer for about 10 minutes, or until it thickens to your favorite consistency. Pour it into a blender and blend until the sauce is smooth.

- If you like chunky barbeque sauce, skip the blender part.
- Pour the sauce in an airtight container and keep it in your fridge for up to 2 weeks. You can also freeze the sauce for up to 3 months.

How to Flavor Cake

Flavoring cake can be an intimidating yet rewarding culinary adventure. There are several ways to alter dull cake mixes and basic cake recipes, so have fun and experiment with a few different options as you search for baked-good-bliss.

Part 1. Adding Flavor to Cake Batter

1. Use extracts and flavorings. Liquid extracts and flavorings are one of the easiest ways to alter the taste of cake batter since they generally have little effect on the other ingredients, moisture level, or density of the cake.

- Extracts are more concentrated than flavorings, so you should generally use less extract than flavoring. Since extracts are more concentrated, however, they typically create richer, more natural tastes.
- To experiment with different extracts, consider starting with a cake recipe that already contains vanilla extract. Replace half of the vanilla with the extract of your choice to introduce a subtle hint of the new flavor.

2. Mix fresh ingredients into the cake. Adding fresh ingredients to the batter can produce an even stronger taste than using the equivalent extract. Some ingredients will blend into the batter, but others will retain their form even after the cake bakes.

- Melted chocolate is a popular choice, but it can be a difficult ingredient to work with. Since the ingredient is so heavy, you will likely need to balance the extra weight with additional beaten eggs or other raising agents.

- Consider scraping out the inside of a vanilla bean instead of using vanilla extract. As a general rule, 2 inches (5 cm) of vanilla bean equals 1 tsp (5 ml) of vanilla extract.
- Chopped ingredients, like nuts, fruits, or large pieces of chocolate can be used and typically retain their form instead of mixing into the batter. Note that fruit can make cakes more moist, however, since the juices tend to seep into the batter as it bakes.

3. Sift cocoa powder into the batter. Cocoa powder is essentially the dehydrated form of cocoa solids. You can add it to white or yellow cakes to create a strong chocolate flavor or mix it into chocolate cakes to further intensify the taste.

- There are two types of unsweetened cocoa powder: Dutch-processed and natural. The former has a lighter flavor and should be used in cake recipes calling for baking powder or other acidic ingredients. The latter can have an intense, bitter taste and should be used in recipes calling for baking soda.

4. Try flavored liquids instead of water. For cake mixes and recipes including water, switch out part or all of the water for another liquid that can introduce more flavor to the cake.

- You can generally swap out half to all of the water when using dairy-based or water-based liquids, including brewed coffee, milk, and juice.

- For alcohol-based liquids, however, you should replace no more than half of the water with your liquor or liqueur of choice.

5. Spice up the batter. Certain spices can add new flavors to the cake or enhance flavors currently used. Some spices are more potent than others, though, so you'll need to vary the amount depending on which one you choose.

- Popular choices include cinnamon, ginger, nutmeg, cloves, and allspice. Of these additions, only cinnamon should be used in relatively large quantities (up to 2 or 3 tsp [10 to 15 ml]). The others tend to be more potent when used in excess.

6. Add dry pudding mix. Dry instant pudding can add both moisture and flavor to a cake, and it's an especially good option to use when you want to flavor a boxed cake mix.

- Typically, you should use one small box of instant pudding in the flavor of your choice for one standard box of dry cake mix. You may also wish to add an extra egg to help balance out the extra weight since eggs can serve as raising agents.

7. Enhance the overall flavor with richer ingredients. If you want to produce a richer taste instead of completing altering the flavor profile, switch blander ingredients for more flavorful options or increase the amount you use.

- Instead of using vegetable oil, use equal amounts of melted butter since butter has a richer taste. Depending on the recipe, you might be able to add as much as twice the amount originally called for, but doing so will also introduce more moisture into the cake.
- Consider cracking an extra egg into the batter. Eggs are another ingredient capable of enhancing the tastes already in your batter, and adding an extra egg can also make cakes fluffier.

Part 2. Adding Flavor to Baked Cake

1. Pour syrup over the cake. As soon as you pull the cake out of your oven, use a toothpick to poke small holes into the surface and pour or spread hot fruit syrup over the entire top.

- The resulting cake will retain the flavor of the syrup while also becoming more moist.
- For a simple fruit syrup, consider mixing 2 to 3 Tbsp (30 to 45 ml) of any fruit jam or preserve with enough hot water to dissolve it into liquid form.

2. Spread filling in between layers. If you baked a layer cake, you can add more flavor to each bite by spreading an interesting filling between each separate layer.

- Choose a filling that can hold its form beneath the weight of the cake's upper layers. Buttercream icing is popular, but jams, canned pie fillings, and canned cake fillings also work well.
- For most fillings, you should spread an even 1/4 inch (6 mm) over the top surface of the bottom layer before placing the next layer over it.

3. Change the frosting. A change in frosting is another easy way to change the overall flavor profile of your cake. You can use prepackaged jars of frosting or mix your own frosting.

- Like cake batter, you can also alter the taste of the frosting using extracts or liquid flavorings. Add the ingredient in small amounts, however, to prevent the frosting from becoming too runny. Moreover, you should also opt for artificial extracts or naturally clear flavorings if you do not wish to tint the final color of the frosting.

Part 3. Picking Flavors

1. Experiment with different combinations. Be bold, and don't be afraid to try flavors and flavor combinations you haven't run across before. Departing from the recipe as it's written requires a little culinary courage, but the end result can be well worth it.

- Look for inspiration everywhere and jot down the ideas as they come. Inspiration often strikes when you eat other foods, learn about unique ingredients, or hear about intriguing cakes. Additionally, your sense of smell is connected to your sense of taste, so don't dismiss ideas derived from candles or other pleasant scents, either.
- Experimenting when the pressure is low can also embolden you. Avoid testing new ideas when baking cakes for special occasions and consider halving the recipe to create less waste if your test fails.

2. Research your idea. If you have an idea but are too scared to take the plunge, do some research before you step into the kitchen. Search on Pinterest or popular recipe websites to determine if anyone else may have combined similar flavors, then read about the methods others have used when doing so.

- Instead of looking for an exact recipe to duplicate, focus on the proportions and ingredients used to produce certain flavors. Doing so can give you an idea of how much to add of which ingredients when you start experimenting with your own cake recipe.

3. Learn about each ingredient. Baking involves chemistry, and many ingredients you might use to add flavor will also affect the cake in other ways. Before using any particular ingredient, find out if it might impact the cake's density, texture, or ability to rise.

- Counteract unwanted effects by adding other ingredients to balance them out. For instance, cocoa powder adds weight and dryness to a cake, so you would need to add more of your raising agent and more moisture to neutralize these effects.
- Note that you may also need to learn about some of basic ingredients; otherwise, you won't know which ingredients to use when balancing out the negative effects of your flavor additions.

4. Build the flavor gradually. Since you'll likely end up working with some unfamiliar ingredients and flavor combinations, it's best to start adding new flavors in small amounts.

- This is especially important when you add flavors directly to the batter. You can easily build up weak flavors, but if you accidentally add too much of any one taste, you won't be able to cut that flavor back down.
- After adding each new flavoring ingredient, taste the batter and determine if it needs more. When adding multiple flavors, ask yourself if the flavors balance each other or if you need to alter the proportions further.

5. Evaluate the finished cake. The baking process can tone down some flavors while drawing out other tastes, so the finished cake may not taste exactly like the batter. For that reason, you'll need to continue evaluating the taste and balance of flavors while biting into the finished cake.

- Be your own critic. Ask yourself if the cake tastes the way you hope it would, and if it doesn't, try to figure out the cause. The flavor combination may not work at all, but oftentimes, the taste can improve with some alterations. It's possible that you added too much or too little of one ingredient. Alternatively, you might decide to swap out one ingredient for another—for instance, melted chocolate for cocoa powder.
- Ask others to critique your cake, as well. Since they didn't bake it, they may not be able to help you alter the recipe for future reference, but they can at least give you an honest opinion about the way it tastes.

How to Flavor Rice

Rice doesn't have to a bland side dish. If you want to spice up rice a little, you have many options. You can cook rice with herbs and vegetables to add flavor during the cooking process. You can also mix in delicious flavors after rice is off the stove. If necessary, adjust the flavor of the rice after you add your spices to make sure it isn't too sweet, spicy, or salty.

Method 1. Flavoring Rice While it Cooks

1. Add salsa to the water for a spicy flavor. If you're serving rice alongside a spicier dish, you can add some spice to the rice to match. Cook rice in a mixture of half salsa and half water for a savory, spicy dish.

- However, keep in mind rice is sometimes used to cool down your mouth when eating something spicy. Use a salsa that's slightly more mild than the dish you're making so the rice isn't overwhelming.
- If you're making a very spicy dish, use something like carrot juice instead of salsa. This way, your rice will have a savory flavor to match without being too hot.

2. Cook rice in broth. Try substituting water for chicken or vegetable broth. This can give your rice a little extra flavor. Chicken and vegetable broth are great flavoring options if you want a side dish with a mild, but noticeable, flavor.

3. Boil rice with diced vegetables or fruits. Vegetables with strong flavors can be diced and cooked

in the pot with rice. Add some diced onions, cranberries, cherries, or peas to the pot where you're cooking rice. The flavor will seep into the rice during the cooking process, creating a rich built-in flavor for your rice.

- You can also add in herbs or spices that pair well with the vegetables you're adding. For example, add in some freshly minced garlic alongside diced onions.

4. Cook your rice with cinnamon sticks and garlic cloves. For a sweet, spicy dessert rice, add a cinnamon stick to your rice before cooking. Add some cloves of cinnamon and garlic as well, for a very strong flavor. Cook your rice as you normally would. When your rice is cooked, you'll have a dish with a bold, slightly sweet flavor.

Method 2. Adding Flavor to Cooked Rice

1. Stir fresh herbs into cooked rice. Fresh herbs can add a powerful flavor to a bowl of freshly cooked rice. After removing your rice from the stove, chop your chosen herbs into tiny slices. Then, mix them into the rice to add flavor. Herbs that can be added to rice include things like parsley, rosemary, cilantro, and garlic.

- Choose herbs that match the dishes you're serving. If you're serving rice alongside an Italian dish, for example, try adding fresh garlic and basil to the rice.

2. Cook your rice in oil after boiling it. After your rice is cooked, heat some oil in a skillet over medium-high heat. Choose an oil that has a lot of flavor, such as olive oil. You can also add some herbs or spices, such as parsley or garlic. Add your rice to the skillet and coat the rice in the oil and herbs until it's golden brown.

- While this option adds a lot of flavor, it also adds extra calories to the rice. It's not the best option if you're looking for a side dish low in calories.

3. Add sweet ingredients for a dessert rice. After you cook your rice, mix in fruits like cranberries and cherries to sweeten it. Add sweet spices, like cinnamon and nutmeg, as well as things like vanilla or almond extract. This can create a sweet-flavored rice that can make a great dessert.

- Flavoring rice with fresh fruit and low or no-calorie sweeteners can be a great dessert option if you're trying to cut down on added sugar.

4. Use commercial seasoning blends on your rice. When you're in the supermarket, pick up

seasoning packets from the spice aisle. Things like cajun blends, or even spice blends meant to flavor taco meat, can be mixed into rice after cooking. This is a very quick means to create a delicious rice side dish for any meal.

5. Add a splash of citrus flavor. Citrus flavors, like lime and lemon, can add some tang to rice. This can work great when serving rice alongside dishes such as fish, which pair well with citrus. After removing rice from the stove, try squeezing in a dash of your favorite citrus flavoring.

Method 3. Avoiding Common Mistakes

1. Make sure to add enough salt. Too much salt can overpower a dish, but don't make the mistake of adding no salt to your rice. Salt, in moderation, can help bring out a dish's flavor. After flavoring your rice, add a small amount of salt and taste it. If the dish tastes a little bland, add salt in small increments until the flavor is at the level you prefer.

2. Adjust the rice if it's too sweet. You can go overboard on sweeteners such as extracts and sweet

spices. If your rice is a little too sweet, you can adjust it by adding a splash of something acidic. A small amount of apple cider vinegar, for example, can tone down a dish's sweetness.

3. Tone down the flavor if your rice is too spicy. Many people enjoy adding a spicy flavor to their rice, but rice is often necessary to cool your mouth between bites of a spicy dish. If you made your rice a little too spicy, tone the flavor down by mixing in vegetables or fruits to dilute the spicy flavor. You can also add a mild amount of dairy, such as sour cream or plain yogurt.

How to Add Flavor to Butter

Adding flavoring to butter can be a quick and simple way to add a lot of flavor for minimal work.

This will gives a range of different flavoring suggestions and as the butter can be frozen, you can make it in advance for use to add to a snack or quick meal, or spread into sandwiches, spread on bread to be grilled or to garnish and flavor meals.

Steps

1. Mix your softened butter with the desired flavoring as per the following suggestions, using a bowl and beaters or a food processor. If you used unsalted butter, you may add a pinch of fine sea salt if desired, but do not use salted butter for sweet variations.

- It's desirable to use fresh rather than dried flavorings as the dried flavors can be stale or musty and never as piquant as fresh ingredients.
- Most butters may be used the same day or immediately after use, however some butters - especially those containing herbs and spices, are best used the next day after the flavors have developed.

2. Place the butter mix onto a sheet of baking paper, foil or cling film and wrap like a you would roll a Sushi roll. Chill in the fridge until needed, of freeze for longer term storage.

3. Slice and serve as desired.

Method 1. Blue Cheese Butter

1. Purchase 30g or a small taster portion of blue cheese. Slice the cheese and butter in thin slices

and put in layers, wrap and chill for 48 hours for the flavors to infuse. You may also blend the cheese into the butter if you wish to use it sooner.

- The butter can be used on its own, or the cheese and butter together may be added to grilled dishes, cheese sauces. Alternatively if made as a cheese & butter paste, used in bread rolls or sandwiches.

Method 2. Garlic Butter

1. Blend butter with 1 tablespoon of finely minced garlic. This butter may also have 2 tablespoons of chopped fresh herbs such as rosemary, parsley, chives, basil, sage or oregano.

- This butter may be used immediately, or chilled for further use. It can taste strongly of garlic, so use according to the guest. For a mellow flavor, roast or gently fry your garlic first.

Method 3. Lime, Chilli and Coriander Butter

1. Mix butter with the freshly grated zest of one lime. Add one de-seeded and chopped fresh red or green chilli (to taste) and 2 tablespoons of freshly chopped coriander / cilantro.

- This butter is delicious on baked or grilled corn on the cob. You may alternatively substitute coriander with chives or 1 minced fresh ginger.

Method 4. Sun-dried Tomato, Basil and Olive Butter

1. Mix butter with 3 whole sun-dried tomatoes finely chopped, as well as approx 6 olives chopped, or enough olives chopped to fill 2 tablespoons. Add 4-5 large leaves of chopped basil.

- This butter has a strong Mediterranean flavor, so can be used to complement any dishes of this theme.

Method 5. Cheese and Pepper Butter

1. Purchase 30 grams of strong tasty cheese, such as Parmesan or Cheddar. Mix the butter with 1 tablespoon of freshly ground black pepper (or add pepper to taste). Finely grate the cheese and add to the pepper butter, alternatively, you may blend cheese and butter in a food processor and add pepper once the cheese is blended. Wrap and chill for 48 hours for the flavors to infuse.

- This butter is delicious with anything that you would like a cheese-like savoury taste, as well as a spicy peppery bite.

Method 6. Mustard Butter

1. Mix butter with 2 tablespoons of mild mustard such as Dijon, Mild English or American mustard. You may alternatively mix 1T hot English mustard, or supplement the mild with a little other strong mustard.

- Mixing the mustard can give a good balance as the butter will make the strength of the mustard seem more mellow. Too much hot English mustard can have an acid taste however, so the key is to balance to your tastes.
- Mustard butter is also good mixed with chopped capers, thyme or chives.

Method 7. Horseradish or Wasabi Butter

1. Mix 2 Tablespoons of horseradish or wasabi with the butter to taste.

- This butter is often served with roast meats, but can also be delectable with steamed vegetables.

Method 8. Sweet Coconut Butter

1. Mix butter with 20 grams of fine desiccated coconut that has been toasted, 1.5 tablespoon of honey or icing sugar and a few drops of coconut essence (to taste).

- This butter is also good with savoury dishes as well as spread on sweet cakes. In savoury dishes its gives a tropical taste.

Method 9. Sweet Citrus Butter

1. Mix butter with 1 & 1/2 tablespoons of sweetener, such as icing sugar, honey, maple syrup, golden syrup or brown sugar. Combine butter with 2 tablespoons of grated zest of lemon, orange, lime, or a combination.

- This is delicious mixed into cakes as you would mix butter normally. Its also nice on hot desserts such as steamed sponge puddings, waffles and pancakes.

Method 10. Sweet Spice Butter

1. Mix butter with 1 teaspoon of cinnamon, a pinch of cloves and / or nutmeg or 1 & 1/2 teaspoon of mixed spice or allspice. Add 1.5 tablespoons of sweetener, such as icing sugar, honey, maple syrup, golden syrup or brown sugar and mix until combined. Ideally leave overnight before use.

- This is delicious spread onto hot baked good such as muffins, toast, cakes, waffles and pancakes.

Method 11. Sweet Nut Butter

1. Use 1/2 a cup of honey or caramel glazed sweet nuts, such as macadamia, almond, hazelnut, pecans etc. Finely chop or pulse in a food processor and combine with the butter. Alternatively, use fresh whole nuts and add 2 T honey or crushed hard caramel sweets.

- This butter may also be used in cakes as a raw ingredient, or used on hot desserts and baked goods.

How to Balance Flavors in your Cooking

Cooking is a skill that takes time to develop. One important aspect of cooking is the art of balancing flavors. Different spices and herbs can create a variety of flavor profiles, so recognize what to include to create the flavor you want. Season food in an effective fashion to get the flavor exactly correct. If you make a mistake, you can adjust the flavor accordingly by adding different ingredients to balance things out.

Method 1. Creating a Flavor Profile

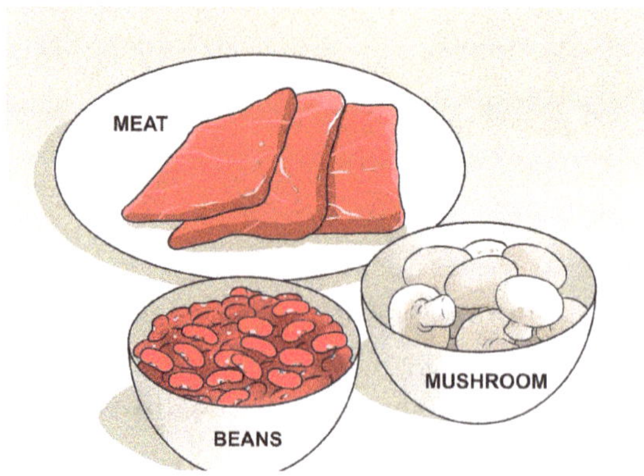

1. Recognize the strengths of different flavors. Different flavors produce different strengths in a dish. Certain types of flavors and spices will be the main thing you taste when eating. Other flavors and spices provide a more subtle taste. Know the general strengths of ingredients to build the right flavor profile.

- Things like meat, mushrooms, and beans provide what are known as low notes. They are the subtle flavors lingering under other tastes in a dish.
- Mid notes are very subtle flavors. While they help inform the taste of a dish, they are not immediately noticeable. Things like raw vegetables and chicken provide mid notes.
- High notes define the flavor of a dish. They are the strongest, most notable flavors. In general, spices, citrus, herbs, and peppers make up high notes.

2. Find ingredients to round out the flavor. Some ingredients round out the flavor in a dish, this

means they bring ingredients together and create a unified taste. You rarely taste these flavors, but they're important to unify a dish. Low notes include things like butter, cream, sugar, and salt.

3. Make a savory profile. For something savory, use the right ingredients to create a strong flavor profile. Savory foods contain ingredients that are saltier and generally not very sweet. Vegetables, meats, and soups can all be cooked with a savory profile.

- Ingredients that round out savory profiles are things like salt, hard cheeses, and fish pastes.
- High notes are things like soy sauce, miso, seaweed, fish sauce, and pickled vegetables.
- Low notes are things like cured meats, such as bacon, tomatoes, and mushrooms.

4. Try a sweet profile. A sweet profile can be used for a number of foods. Many side dishes, like sweet potatoes, are sweetened before being served. Opt for the right mix of flavors to create a sweet profile.

- Honey, jam, sugar, and maple syrup are used to round out a sweet flavor profile.
- Low and mid-notes for a sweet profile include mostly vegetables. Things like sweet potatoes, corn, carrots, and squash work here. Most fruit can also be used to provide low or mid notes for a sweet profile.

- For your high notes, molasses, ketchup, and barbecue sauce can be used. Different types of vinegar, like balsamic and apple cider vinegar, also work well.

5. Add something sour. In addition to a flavor profile on its own, adding a splash of sour flavor can help round out sweet and savory foods served together. For example, sweet and sour chicken helps blend the sweet sauce with the savory chicken.

- Round out a sour flavor profile with ingredients like lemon, lime, or orange juice. You can also use vinegars such as sherry, red, rice, balsamic, and apple cider vinegar.
- Tomatoes can provide mid or high notes for a sour profile.
- For high notes, try ingredients like tomato paste, yogurt, sour cream, and pickled vegetables.

6. Create something bitter. Bitter flavors are not used often for main meals, but you may want to add a mild bitter flavor to balance out something very sweet or savory. Bitter flavors may also be used in desserts, dips, or side dishes. Bitter flavors generally do not have a mix of high or low notes and are only tossed in to balance out other flavors.

- Things like coffee, cocoa powder, grapefruit juice, or beer create a bitter flavor.
- Green vegetables, like kale, spinach, and broccoli also provide a touch of bitterness.

7. Make spicy food. Spice is generally used in sweet, savory, or sour flavor profiles. It can work well to balance out sour and sweet foods. You can add spice by throwing things like hot sauce, wasabi, horse radish, or spicy peppers into a dish. Food like arugula and radishes also have a spicy flavor.

Method 2. Seasoning Food Effectively

1. Balance flavor profiles across ingredients and side dishes. In addition to seasoning individual ingredients, make sure you use complimentary flavor profiles between dishes. Your side dishes and main courses should have a more or less even flavor profile.

- For example, spicy, savory curry goes well with mild soup or rice.
- If you're serving dishes in courses, a sweeter salad pairs well with a sour entree.

2. Add enough salt. Salt is vital to the cooking process. You should not be able to taste salt in a

dish. In fact, if a dish tastes salty this is a sign you've added too much salt. A dish, however, should be lightly sprinkled with salt to bring out other flavors. After adding your flavor profile, sprinkle a little bit of salt at a time until the flavors are strong enough.

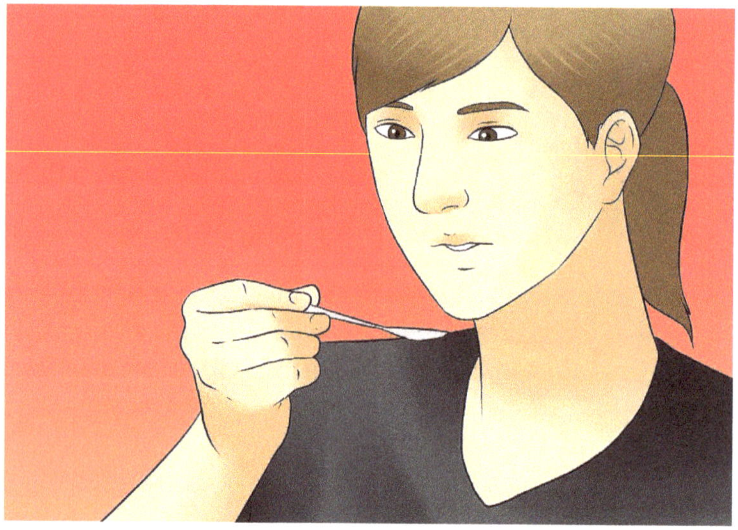

3. Cleanse your palette between tastes. When you're seasoning food, you will be unable to taste your flavors if you do not cleanse your palette. The more you sample food, the less you can taste it. Between taking bites to gauge taste, sip on water or something else with a neutral flavor. This will help you keep your taste buds alert to the precise flavors in your dishes.

4. Season food at the serving temperature. Always season your food at the temperature you intend to serve it. Hotter temperatures enhance flavors, while cooler temperatures diminish them. It is hard to tell the right flavor when the dish is at a higher or lower temperature than you intend to serve it.

- If you reheat leftovers, you usually have to salt them to taste afterwards to get the right flavor profile back.

Method 3. Making Flavor Adjustments

1. Add salt if a dish is bitter. A bitter dish can be adjusted with salt. If you added too many bitter ingredients, like bitter vegetables or alcohol-based seasoning, sprinkle some salt over the dish. Add enough salt that the dish loses its bitter flavor.

2. Adjust overly sweet food with bitter or sour ingredients. If you ended with an overly sweet flavor profile, add bitter or sour ingredients to balance this out. A small splash of something like, say, lemon juice can balance out a sweet dish.

- Remember, bitter ingredients include things like coffee, beer, and green vegetables. Sour foods include things like yogurt, sour cream, and vinegars.

3. Add sour ingredients wisely. Sour ingredients should be used in small amounts, as they can easily overwhelm a dish. A small squeeze of juice from a lemon, for example, will go a long way. In general, sour is best used to balance out sweet flavors than as a profile on its own.

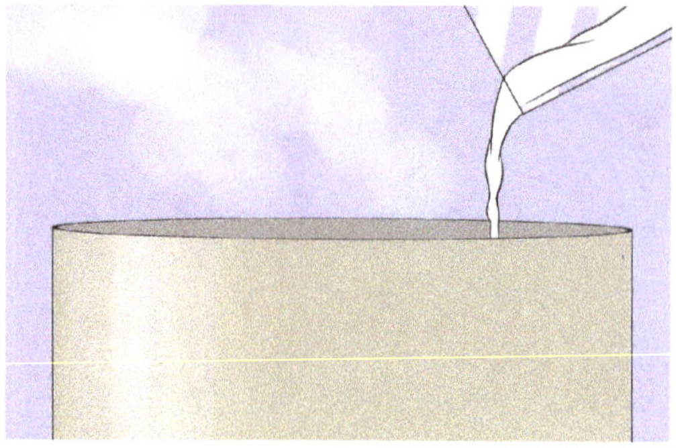

4. Use water to fix an overly salted dish. If you can taste the salt in a dish, you've added too much. Luckily, adding a little water can dilute an overly salted dish.

- If you don't mind a little sweetness, you can also add a sweet ingredient like honey to balance out an overly salty dish. However, you should only go this route if you want to add another flavor to your dish.

How to Intensify the Flavors of your Turkey

In a way, a turkey is like a blank canvas, waiting to take on whatever palette of flavors you hand it. To create a really vibrant package, we take a three-pronged approach to my bird, making sure that each of the added components is truly packed with flavor. We start by rubbing the skin and inside cavity with sage, nutmeg, and pepper, an aromatic combination that transfers its volatile oils to the bird while releasing an uplifting fragrance into the kitchen. For example, my sausage stuffing doesn't stop at bread, sausage, and the usual lineup of aromatic vegetables. For deeper flavor, add fresh thyme leaves, Marsala wine, and the turkey liver. And instead of making a gravy of just pan drippings, roux, and stock, enrich the sauce even further by whisking in Madeira or fresh herbs. These seemingly minor touches help elevate the turkey—and the whole meal—to something memorable and delicious, yet still traditional.

Choose the Bird and Season it Well

Turkeys these days sport all sorts of confusing monikers on the label: self-basting, free range, or all natural. Always look for the word natural, which means the bird has been only minimally processed, in contrast to self-basting birds, which have been injected with oil or water. Freerange birds must be ordered ahead from natural food stores and some groceries, but in my experience they have been leaner and drier than the more readily available “natural” birds.

One more note about label reading: look for the words “includes giblets.” You’ll need those for making gravy.

As for size, generally prefer smaller birds, mostly because larger ones won’t fit in the refrigerator. The rule of thumb is about one pound of turkey per person; if you want substantial seconds and leftovers, allow another half-pound per person. If you’re hosting a really big crowd and you have two ovens, consider roasting two smaller birds instead of one large one. (This will also provide a good excuse to try two kinds of dressing.)

A roasting pan made of a heavyweight metal helps ensure that the drippings don’t burn while the turkey roasts, and it gives better results when it comes time to deglaze the pan for making gravy. But if you don’t have one, don’t despair; the disposable aluminum ones will work.

Be liberal with the seasoning. That the most common mistake cooks make when roasting a turkey is using too little salt and pepper. After rinsing the bird thoroughly and patting it dry, rub it with melted butter or olive oil and season it with plenty of coarse salt, freshly ground black pepper, nutmeg, and dried sage. Sweet-smelling nutmeg amplifies the aromatic qualities of everything else, and sage seems to bring out the homey goodness of poultry. The warmth of thyme is a viable option as well, and since my grandmother is allergic to sage, this is a frequent substitution in my family.

Choose a good-quality bread, and buy it a couple of days in advance. French bread comes in all guises, from mediocre supermarket loaves that are empty of flavor to excellent baguettes with crackly crusts and an irregular crumb. Obviously, go for the latter, if possible.

Thanksgiving is one event where it’s okay to buy the bread a few days early because you want dry, stale bread for stuffing so it can stand up to the added moisture without getting mushy, as fresh bread would. You could also cube fresh bread several hours before making the stuffing and let the cubes air-dry at room temperature, or toast them briefly in a low oven.

If you opt for cornbread stuffing, make your own cornbread a day or two ahead. You can whip it up in about a half hour. Cornbread crumbles easily, so cut it up into fairly large cubes, which naturally crumble into smaller bits during mixing.

For springier breads that don’t crumble, cut smaller cubes. When the cubes are too large, the dressing feels more like a hodgepodge of disparate ingredients and less like an ensemble. This is especially the case for dry stuffings that lack a binding element, like eggs.

These recipes make a lot of stuffing because there never seems to be enough of it. About a third of the dressing will fit inside a 12- to 14-lb. bird, and the rest can be cooked in a baking dish. Don’t shovel in the dressing as if you are filling a sausage casing, also don’t worry excessively about overpacking.

If you want to get a head start on the stuffing, you can cut up the bread, sauté the vegetables, and measure the seasonings in advance. To avoid soggy bread and spoilage, don’t combine the bread with the liquid components until you're ready to stuff the bird. This is particularly important for dressings that are bound with eggs, like the cornbread stuffing.

Stuff both cavities

Gravy: The Final Step Before Carving

No turkey dinner is complete without gravy. And for me, no turkey gravy is complete without a rich base of stock made from the giblets. The stock is so simple to make, and the timing so well coordinated with roasting the turkey.

As flavorful as it is, stock is too light to stand alone as a base for gravy. For real intensity of flavor, you need to incorporate the pan drippings, those dark, caramelized juices on the bottom of the roasting pan. For really smooth gravy that isn't gummy, use a judicious hand with any thickeners and whisk constantly to avoid lumps.

One Way to Thicken—A Butter-flour Paste

To thicken with a beurre manié (pronounced burr mahn-yay), knead equal parts soft butter and flour into a smooth paste.

Bring strained giblet stock and defatted, deglazed pan drippings to a boil and whisk in the beurre manié a bit at a time.

Another Starts with a Fat and Flour Roux

To thicken with a roux (pronounced roo), put the fat reserved from the pan drippings in a pan over medium heat. Whisk in the flour.

When the roux just begins to turn blond, whisk in strained giblet stock and defatted, deglazed pan drippings.

Chapter 3

Seasoning: A Process for Improving Flavor

Seasoning is a technique in food preparation in which salt, herbs or spices are added to food to enhance its flavor. Other seasonings like sea salt, black pepper, basil and seasoning sauce can also be used. This chapter examines the various processes of improving flavor through an understanding of different seasonings and seasoning techniques.

Seasoning is the process of adding salt, herbs, or spices to food to enhance the flavor.

Seasonings include herbs and spices, which are themselves frequently referred to as "seasonings". However, Larousse Gastronomique states that "to season and to flavor are not the same thing", insisting that seasoning includes a large or small amount of salt being added to a preparation. Salt may be used to draw out water, or to magnify a natural flavor of a food making it richer or more delicate, depending on the dish. This type of procedure is akin to curing. For instance, sea salt (a coarser-grained salt) is rubbed into chicken, lamb, and beef to tenderize the meat and improve flavor. Other seasonings like black pepper and basil transfer some of their flavor to the food. A well designed dish may combine seasonings that complement each other.

In addition to the choice of herbs and seasoning, the timing of when flavors are added will affect the food that is being cooked.

In some cultures, meat may be seasoned by pouring seasoning sauce over the dish at the table. A variety of seasoning techniques exist in various cultures.

Infused oils are also used for seasoning. There are two methods for doing an infusion—hot and cold. Olive oil makes a good infusion base for some herbs, but tends to go rancid more quickly than other oils. Infused oils should be kept refrigerated.

How to Make Poultry Seasoning

For many people, poultry seasoning stays in the cupboard until Thanksgiving. But there's no reason this flavorful and versatile seasoning should only be used once a year. Traditional dried poultry seasonings will last several years while fresh herb poultry seasonings can add intense fresh flavor to your dishes. Learn how to incorporate any poultry seasoning into your daily cooking for a quick flavor boost.

Method 1. Making a Traditional Poultry Seasoning

1. Get your supplies and ingredients. Gather a small bowl, a storage jar or container for your poultry seasoning, a small whisk, and a set of measuring spoons. This traditional dried poultry seasoning recipe will make about 1/4 cup. Most recipes calling for poultry seasoning will only require a teaspoon or two, so you'll make more than you might need with this recipe.

- If your spices are dried or lumpy, you may need to replace them with fresher spices. Taste and smell the spices. If they're still flavorful, you may just want to sift them.

2. Mix the dried poultry seasoning. Place all of your dried ingredients into the small bowl and use the whisk to thoroughly combine the seasoning. Once the spices are combined, your poultry seasoning is ready to store or use. You'll need to measure out and combine:

- 1 tablespoon of dried ground sage

- 1 tablespoon of dried ground thyme
- 1 tablespoon of dried ground marjoram
- 1 teaspoon of dried ground rosemary
- 1 teaspoon of crushed celery seed
- 1 teaspoon of ground black pepper

3. Use your dried poultry seasoning. Poultry seasoning is famously rubbed on turkey before roasting it. You could also rub it on chicken or capons before roasting them. If using the traditional dried seasoning, drizzle a little olive oil over the meat and sprinkle 1 or 2 teaspoons of the dried seasoning over the poultry. Use your hands to evenly rub the seasoning across the entire turkey or chicken. Cook the poultry according to your specific recipe.

- You may need more or less seasoning, depending on the size of your turkey or chicken.

4. Store the dried seasoning. Cover the seasoning with a tight-fitting lid or transfer your dried seasoning to an airtight container. Spices and seasonings can last several years if stored away from heat and light. But you should still try to make poultry seasoning in small amounts so you use it quickly.

- If you'd like your poultry seasoning to stay even fresher for longer, consider purchasing whole spices to grind yourself.

Method 2. Making Fresh Herb Poultry Seasoning

1. Get your spices, herbs, and supplies. You'll need a small bowl that holds about 8 ounces, a storage jar or container for your poultry seasoning, a spoon or whisk, a set of measuring spoons, a sharp knife, and a cutting board. This fresh herb poultry seasoning will make around 3/4 of a cup.

- Since the flavor of fresh herbs isn't as concentrated as in dried herbs, you'll need to use more of the fresh seasoning than you would of dried poultry seasoning.

2. Chop the fresh herbs. Rinse the herbs and remove the stems since they can be woody or tough. Place them on a cutting board. Carefully chop the fresh herbs using a sharp knife until they are fine. You'll need to chop:

- 3 tablespoons of thyme
- 3 tablespoons of rosemary
- 3 tablespoons of sage
- 1 tablespoons of marjoram

3. Mix the fresh poultry seasoning. Place the fresh herbs in the small bowl along with the other spices. Use the spoon or whisk to combine the seasoning. Your seasoning is now ready to store or use. You'll need to combine the fresh chopped herbs with:

- 1/2 tablespoon of freshly ground black pepper
- 1 teaspoon of celery seed
- 1 teaspoon of freshly grated nutmeg
- 1 teaspoon of ground ginger
- 1 teaspoon of smoked paprika

4. Store the fresh poultry seasoning. Cover the seasoning with a tight-fitting lid or transfer your fresh herb seasoning to an airtight container. Since the herbs are fresh, you'll need to use them quickly or they'll begin to lose their flavor. Try to use this seasoning as soon as you make it.

- If you need to make the seasoning ahead of time, you can refrigerate it up to 3 days.

5. Use your fresh herb seasoning. Drizzle a turkey or large chicken with a tablespoon of olive oil. Apply all of the fresh herb seasoning and rub it all over the meat using your hands. You should have a heavy coating of the seasoning evenly distributed across the poultry. Cook the turkey or chicken according to your specific recipe.

- You'll need to use all of the seasoning since fresh herbs aren't as concentrated in flavor as dried herbs.
- You can also use the seasoning when cooking individual pieces of poultry.

How to Make Taco Seasoning Mix

Keep a package of homemade taco mix stashed away in the cupboard for a quick and inexpensive meal. Just say hey and dinner is on the table. Serve tacos with favorite condiments such as shredded lettuce, grated cheddar cheese, chopped tomatoes and favorite salsa.

Method 1. Making the Seasoning

1. Mix all the ingredients of either variation in a small bowl thoroughly. Alternately, shake in a small jar to combine. Each variation has its own distinct flavor. Experiment with what tastes best for you. Each recipe contains chili powder, cumin, and paprika, but variation #1 has coriander and cayenne, while variation #2 has onion and garlic powder, in addition to red pepper flakes and oregano.

2. Store taco seasoning in an airtight container for up to several months. Use sooner rather than later for best results.

How to Make Ranch Seasoning

Love the ranch seasoning that you get at the store, but don't love the chemicals? Try making your own.

Steps

1. Put all of the ingredients into a small to medium sized bowl.

2. Whisk them all together.

- Use a food processor to make a finer seasoning. Pulse it for a few seconds.

3. Store it in an airtight container.

4. Note when the seasoning is made. It can be kept fresh up to 3 months.

How to Make Jerk Seasoning

Jerk seasoning is an aromatic rub or marinade used in Caribbean cooking. There are several variations on traditional jerk seasonings and marinades, ranging from very spicy to mild. You can use this sauce to make flavorful chicken, pork, fish, and seafood dishes.

Method 1. Preparing Spicy Jerk Marinade

1. Consider making a spicy jerk marinade. Traditionally, jerk marinades combine many spices such as pepper, nutmeg, cinnamon, and allspice with hot peppers, onions and garlic.

- This marinade is an excellent flavorful addition to grilled chicken and other meats.
- Keep in mind that jerk seasoning is meant to be spicy.
- The scotch bonnet pepper is traditionally used in authentic Jamaican cuisine. These can have quite a bit of heat.

2. Measure out the spices. You will need a collection of spices for this marinade. Keep in mind this is a highly flavorful and spicy sauce.

- You will need 2 teaspoons salt, 1/2 teaspoon black pepper, 1 tablespoon brown sugar, 2 teaspoons ground allspice, 1 teaspoon nutmeg, and 1 teaspoon cinnamon. For the fresh spices and seasonings you will need 5 sprigs of fresh thyme and 1 teaspoon grated fresh ginger.
- Combine all the spices in a small bowl.
- Set these aside while you prepare the other ingredients.
- It is a good idea to measure out spices and mix them before preparing other ingredients.
- All ingredients will need to be blended thoroughly. Having spices mixed ahead of time will make this process go faster.

3. Peel onions and garlic. Chop these finely.

- Set these aside in a small bowl.
- You will eventually add these to a blender to make a thick marinade.
- You can always chop garlic and onions ahead of time and store these in the refrigerator until you need to use them.

4. Prepare the scotch bonnet peppers. Take care while working with these, as they are very spicy.

- A good rule of thumb when working with hot peppers is to wear gloves.
- If you get the hot pepper juice on your skin and then touch your eye, you could end up with a burning sensation.
- Slice the peppers open and remove the seeds.
- Finely dice the peppers and put them aside in a bowl.
- If you get hot pepper juice on your skin, wash your hands thoroughly with dish soap. Alternatively, you can rub your hands with rubbing alcohol or olive oil. These products will help to break up the capsaicin that makes peppers spicy.

5. Add all of the ingredients to a blender. You will need to use an electric blender to get a thick, smooth marinade.

- Blend the ingredients well, using a higher setting.
- If you find that your blender isn't mixing all of the ingredients, you may need to pause and scrape the sides of the blender.
- If you are making a large amount of marinade and don't have a high capacity blender, you can use a food processor for this.

6. Use the marinade to season meats and poultry.

- Jerk seasoning is best when it is used shortly after you make it.
- Jerk chicken is one of the most common dishes made with this marinade, but you can also use the sauce on pork, fish, or seafood.
- When using jerk sauce as a marinade, make sure you marinade your meat at least 2 hours.
- For a more potent flavor, marinade your meat overnight.

Method 2. Making a Mild Jerk Marinade

1. Gather all of the spices and mix them. You will need to combine the salt, sugar, allspice, nutmeg, cinnamon and black pepper.

- Measure each spice out and put it in a bowl or jar.
- If you want, you can make this spice mix ahead of time and store it for when you want to make jerk marinade.
- You will need to add this spice mixture to a blender later to make a milder version of jerk sauce.
- By mixing the spices ahead of time, you will save time when going to blend the ingredients.

2. Combine the onions, garlic, and thyme sprigs. Chop these ingredients finely.

- Set these aside in a bowl.
- These can be chopped ahead of time to save time.
- If you do so, store these in the refrigerator in a bowl with plastic wrap.

3. Prepare the hot pepper. Take care when working with hot peppers.

- You can use gloves if you want to avoid getting any juice on your skin.
- Make sure you use caution when cutting the pepper, as any juice can cause eye and skin irritation.
- Remove the seeds from the pepper and discard them.
- Pepper seeds are very hot. They can increase the heat in a dish.
- Cut the pepper into small, fine pieces.
- If you get any pepper juice on your hands, be sure to wash them thoroughly with dish soap and water.

4. Combine all ingredients in a blender jar. Blend the mixture on high until smooth.
   - If some of the ingredients stick to the side of the blender jar, pause and scrape the sides.
   - Store the sauce in a tightly sealed jar.
   - This can keep for up to a week in the refrigerator.
   - You can also use a food processor to blend the ingredients.

5. Use the sauce as a marinade. This is best on chicken, pork, or fish.
   - Allow at least two hours for your meat to marinade.
   - Bake or grill your dish for best results.
   - Traditionally, jerk chicken is often grilled. You can even grill a pork loin or pork chops.
   - Be aware that even though this version of jerk marinade is milder than traditional jerk seasoning, it still contains hot peppers.
   - If you find your dish is still too spicy, you can top it with yogurt or sour cream to cut the heat.

How to Make Italian Herb Seasoning

Italian herb seasoning is tasty and suitable for seasoning a wide range of dishes from pastas to stews. This seasoning tastes best if you have dried the herbs yourself; that way it is guaranteed to taste the freshest and is so much nicer than buying it ready-made off the shelf. This recipe is a base version; there are numerous variations on this one and you are free to add more variants as you please.

Steps

1. Mix together in a small bowl.

2. Funnel the ingredients into a small bottle with a lid.

3. Seal and store for up to six months.

How to Make Lemon Pepper Seasoning

Lemon pepper seasoning can be added to a wide range of dishes, and preparing your own doesn't take much effort. You can make quick lemon pepper seasoning with dried spices, but making it with fresh ingredients will result in an even stronger flavor.

Method 1. Quick-Mix Lemon Pepper Seasoning

1. Combine the ingredients. Place the dried lemon peel, ground black pepper, and table salt in a bowl. Stir the three ingredients together until evenly and thoroughly combined.

- Since all three spices are already in finely ground forms, they should mix together without difficulty.
- You can skip the salt or reduce the amount if you'd prefer your seasoning blend to have less sodium. You can also change the amount of lemon peel and pepper used, but these proportions will result in a seasoning blend similar to commercially prepared versions.

2. Store in an airtight container. Pour the lemon pepper seasoning into an airtight jar or similar container. Store it in a cool, dry location for up to six months.

Method 2. Shortcut Lemon Pepper Seasoning

1. Place the ingredients in a food processor. Add the dried lemon, peppercorns, and salt to the bowl of a food processor. Quickly stir them together with a fork.

- Use a spice grinder or coffee grinder if you do not have a food processor.
- Since the dried spices used in this version are larger in size, you'll need to grind them down into finer pieces before you can use them together as a spice blend.
- If you need to watch your salt intake, you can reduce or omit the salt used in this recipe. You can also adjust the amount of lemon and pepper as desired.

2. Pulse until finely ground. Process the mixture in several short bursts, continuing until everything is finely ground and evenly combined.

- Pause to stir the mixture as needed if some of the spices do not reach the blades on their own.
- Pulsing the spices in this manner should grind them down while simultaneously mixing them together.

2. Pulse until finely ground. Process the mixture in several short bursts, continuing until everything is finely ground and evenly combined.

- Pause to stir the mixture as needed if some of the spices do not reach the blades on their own.
- Pulsing the spices in this manner should grind them down while simultaneously mixing them together.

3. Store in an airtight container. Scoop the lemon pepper seasoning into an airtight jar. Store the blend in a cool, dry area for up to six months.

Method 3. Fresh Lemon Pepper Seasoning

1. Preheat the oven to 200 degrees Fahrenheit (90 degrees Celsius). Prepare a baking sheet by lining it with parchment paper.

- If your oven does not reach temperatures as low as 200 degrees Fahrenheit (90 degrees Celsius), simply preheat it to the lowest temperature possible.

2. Zest the lemons. Grate four to six lemons using a standard kitchen grater. You'll need about 3/4 cup (180 ml) of fresh lemon zest.

- Start with four lemons, then zest more if they do not produce enough zest. You can also use more or less than 3/4 cup (180 ml) if you'd prefer a stronger or weaker lemon taste.
- Before zesting the lemons, rinse them under running water and dry with clean paper towels. Only use clean, unblemished lemons for this recipe.

3. Crush the peppercorns. Place the peppercorns in a spice grinder and pulse for several seconds. The peppercorns should be lightly and consistently crushed.

- Do not fully grind the peppercorns during this step.
- If you do not have a spice grinder, use a food processor or clean coffee grinder
- For a simple variation, consider combining black peppercorns with green or pink peppercorns in equal parts. You can also use a "peppercorn medley" that contains a premixed amount of all three types.

4. Combine the lemon zest and peppercorns. Place the lemon zest and crushed peppercorns in a mixing bowl, then stir them together until evenly combined.

- While stirring, use the back of a fork to mash the lemon zest and crushed pepper together. Doing so should help hold the mixture together, thereby blending the flavors more effectively.

5. Bake until dried. Transfer the mixture to your prepared baking sheet, then place it in your preheated oven. Bake it for 60 minutes, or until all of the zest has completely dried.

- After pouring the mixture onto the center of your parchment-lined baking sheet, use a spatula to spread it into an even layer.
- Check the mixture after the first 30 minutes and gently stir it before letting it bake longer. Continue checking every 10 to 15 minutes after that point, stirring as needed to ensure even drying.

6. Grind down the mixture. Scoop the dried mixture back into the spice grinder and process it until everything is finely ground.

- As before, a food processor or coffee grinder can be used if you do not have a spice grinder.
- The finished texture should resemble commercially prepared lemon pepper blends, which is approximately the same texture as fine sand.

7. Combine with salt. Dump the lemon pepper blend into a mixing bowl and add the salt. Stir well until the ingredients are evenly combined.

- If you choose to use large-grain salt, you may wish to place it into the spice grinder with the lemon pepper blend and pulse it into the mixture. This should mix the salt into the seasoning while simultaneously breaking it into finer pieces.
- You can technically prepare this mixture without the addition of salt, but it may not taste like most commercially prepared versions if you leave it out.

8. Store in an airtight container. Transfer the lemon pepper seasoning to an airtight plastic or glass container. Store it in a cool, dry place for up to six months.

How to Make Seasoned Pepper

Pepper is already a tasty seasoning in its own right––with piquant essential oils and the crystalline substance, piperine. When other seasonings are added to pepper though, it becomes an all-round seasoning that can be used on a range of foods to make them even more delicious. The mixes provided here are ideal for flavoring soups, stir-fries, stews, pies, dips, and more.

Method 1. Black and White Pepper Mix

1. Place all of the ingredients into a mixing bowl. Mix together well.

2. Tip into a storage container. An airtight glass spice bottle is a great choice.

3. Use as needed. Add to dishes to spice them up.

- Use within 3 to 6 months, for the best flavor.

Method 2. Lemon Pepper

1. Wash the lemons to remove grime. Peel the lemons with a microplane or zester.

2. Spread the peel pieces across a baking sheet. Place in the oven, middle rack, and heat at the lowest level it has.

3. Bake at this low temperature until the zest feels dry when you touch it. Check the dryness every half hour or so.

4. Crack the peppercorns using a pepper mill, grinder or mortar and pestle. Make enough for 2 teaspoons.

5. Grind or powder 1 tablespoon (14.8 ml) of dried lemon zest. Use the grinder or mortar. Don't make it too powdered; it needs to have some texture.

6. Mix the ingredients together. Add the cracked black pepper, ground lemon zest, sea salt, granulated or powdered garlic and onion powder to a mixing bowl. Stir to combine.

7. Transfer to a shaker container. Store airtight. Use within 3 to 6 months.

How to Make Mustard, Herb and Spice Seasoning

An excellent substitute for salt, this seasoning can also be used to add flavor to a range of dishes from stews to marinades.

Steps

1. Combine all of the ingredients together in a mixing bowl. Mix together well.

2. Pour into a suitable shaking container, such as a cleaned herb jar or similar. Label and date.

3. Use. This seasoning can be used as a salt replacement in any recipe that requires salt. It can also be sprinkled over foods that you'd normally sprinkle salt over.

How to do Seasoning Process for Improving Flavor

1. Drop (Salt for) Acid

In addition to grabbing the saltshaker to boost flavor in soups, stews, and sauces, try a drop of lemon juice or vinegar. Like salt, acid competes with bitter flavor compounds, reducing our perception of them as they "brighten" other flavors. Just a dash—1/8 teaspoon—can go a long way.

2. Use Coarse Salt when Seasoning Meat

Use kosher salt—rather than table salt—when seasoning meat. Its larger grains distribute more easily and cling well to the meat's surface. When a recipe calls for seasoning meat "to taste," we suggest using about 1/8 teaspoon of kosher salt per portion.

3. Pep up—or Tone Down—your Pepper

When exactly you apply black pepper to meat—before or after searing—will affect the strength of its bite. If you want assertive pepper flavor, season meat after searing; keeping the pepper away from heat will preserve its volatile compounds. Alternatively, seasoning before cooking will tame pepper's punch.

4. Season Cold Foods Aggressively

Chilling foods dulls their flavors and aromas, so it's important to compensate by seasoning generously—but judiciously. To keep from overdoing it, season with a normal amount of salt before chilling and then taste and add more salt as desired just before serving.

5. Incorporate Fresh Herbs at the Right Time

Add hearty herbs like thyme, rosemary, oregano, sage, and marjoram to dishes early on in the cooking process; this way, they release maximum flavor while ensuring that their texture will be less intrusive. Save delicate herbs like parsley, cilantro, tarragon, chives, and basil for the last minute, lest they lose their fresh flavor and bright color.

6. Add a Little Umami

Common pantry staples like soy sauce, Worcestershire sauce, and anchovies contain high levels of glutamates that can give a savory umami boost to a dish. Try mixing a teaspoon or two of soy sauce into chili or adding a couple of finely minced anchovies to a chicken braise.

7. Make Adjustments when Seasonings Go Awry

If you've added too much salt, sugar, or spice to a dish, the damage is usually done. In mild cases, however, the overpowering ingredient can sometimes be masked by the addition of another from the opposite end of the flavor spectrum. Consult the following tips for ideas. And remember to account for the reduction of liquids when seasoning a dish—a perfectly seasoned stew will likely taste too salty after several hours of simmering. Your best bet is to season with a light hand during the cooking process and then adjust the seasoning just before serving.

- If your food is too salty, add an acid or sweetener such as vinegar; lemon or lime juice; canned, unsalted tomatoes; sugar, honey, or maple syrup.
- If your food is too sweet, add an acid or seasonings such as vinegar or citrus juice; chopped fresh herb; dash of cayenne; or, for sweet dishes, a bit of liqueur or espresso powder.
- If your food is too spicy or acidic, add a fat or sweetener such as butter, cream, sour cream, cheese, or olive oil; sugar, honey, or maple syrup.

8. Add a Finishing Touch

Even the most perfectly cooked soup, stew, or pasta dish can benefit from a last-minute burst of flavor. One of our favorite ways to liven up rich lasagnas or hearty braises is to sprinkle them with the classic Mediterranean garnish known as gremolata. This mixture features minced fresh garlic, citrus zest, and fresh herbs such as parsley or basil. Soups, pasta, fish, and just about any cut of meat will benefit from a dollop of herb butter made from blending finely minced herbs, garlic, and often shallot into softened butter.

How to use Herbs to Add Flavor to Food

Herbs are jam-packed with antioxidants, vitamins, minerals, phyto-nutrients, and of course, they're sodium-free so they're better for your heart and blood pressure than that nasty salt shaker.

Of course, a little bit of high-quality sea salt is good for you on occasion, but for the most part, herbs are all you need to make a meal nutritious and satisfying. Here's some tricks to get you started:

1. Add them to Smoothies

One of the best ways to add a little flavor to your smoothies is to add herbs like basil, mint, parsley, or cilantro to your smoothies. Just one or two sprigs is all you need. Basil and mint pair beautifully with strawberries, while parsley and cilantro can replace half of the other greens you usually use in your green smoothies. When in doubt, choose mint, which goes great with just about anything, especially raw cacao, berries, spinach, and melon.

2. Sauté with Oil

Before cooking your meals, add a little bit of olive oil or coconut oil to the pan and a couple sprigs of herbs like oregano, basil, thyme, rosemary, or even lavender for a fun twist. Saute the herbs lightly for just about 3-4 four minutes on low heat until the oil becomes fragrant. Remove the herbs and toss them or save for later. Your oil is now flavored and you'll get those same delicious flavors into whatever veggies or grains you put into the pan to cook with the oil.

For instance, saute some oregano, basil, and rosemary with some olive oil for just about three minutes. Then remove the herbs, saute some onions and garlic into the pan with the oil, and then add some cooked quinoa, juice from a lemon, and finally some shredded kale for a smashing dinner dish that's full of flavor and healthy for you too.

3. Mince them First

Herbs truly don't taste that great unless in their whole form, especially in raw dishes like salads, so either mince them beforehand, or shred them with your fingers into tiny pieces. Mincing (or chopping) breaks down the herbs and releases their flavor while also reducing any bitterness or pungency they may have. This trick is especially important for herbs like cilantro, sage, and parsley which are delicious, but carry a strong flavor profile that can be overwhelming in their whole form.

4. Add an Acidic Medium

Herbs used alone in dishes can carry an "off" taste unless you have other ingredients that compliment them and help to balance them out. One of the best ways to do this is with an acidic medium, such as lemon juice, orange juice, apple cider vinegar, balsamic vinegar, berries, or tomatoes. It's one reason most dishes with herbs always include one of those ingredients or some other type of acidic medium. In fact, not only does this tip make the herbs taste better but the mixture of alkaline herbs and healthy acidic mediums also help balance your body's natural pH for optimal health. This is especially a great idea to implement into vegan salad dressings, so give it a try.

5. Don't Add too Many

Because herbs are so tiny, most of us tend to add too many to one dish. This is a big no-no for flavor because of how powerful herbs can be. A little bit goes a long way, and while basil is great on a sprouted grain sandwich or a nice strawberry quinoa salad, you don't want your entire meal tasting like basil. Just one little leaf is all you need. Remember with herbs, less is always more.

Chapter

4 Food Coloring

Food coloring is any substance that imparts color to a food or drink. They can be in the form of liquids, gels, pastes and powders. Color additives are added to food to improve its visual appeal, enhance natural colors or offset color loss. The topics elucidated in this chapter on food coloring cover the various aspects of making natural food coloring, black food coloring, brown food coloring, vegetable dyes for Easter Eggs, etc.

Food coloring is a type of food additive which imparts color. Some such additives are dyes, meaning that they are water soluble, and used in a variety of foods which contain liquids. Lakes have a dye base, but they are designed to disperse in oil as well as water, and they can be used in preparations which lack sufficient moisture to hold dye. Lakes tend to be less prone to coating the mouth, a common problem seen with dyes used in things like frozen treats.

People have been adding color to their food for thousands of years. Early food colorings were commonly made from spices such as turmeric, along with crushed seeds. People added color to food historically for many of the same reasons that we do today: to enhance natural colors, to lend a more familiar color to food, and for decorative purposes. Food coloring is especially popular with preserved foods, which tend to turn dull gray. Although these foods are perfectly safe to eat, they don't look quite right, and coloring can make them more appetizing.

Other sources of natural coloring for food include some minerals, along with several members of the insect world. Cochineal red, for example, comes from beetles. Chemists have also developed purely synthetic food coloring, to obtain especially bright colors or to create colors which are hard to find in nature, such as blue.

Like other food additives, food dyes must be tested for safety before it can be used. Several colors once recognized as safe have since been removed from the market due to health concerns, and some consumers try to avoid all artificial colors, even those which have been tested and approved. Avoiding food coloring can actually be quite difficult, as one of its most common uses is situations where natural colors need to be restored. Many oranges, for instance, are sprayed with coloring to cover up their naturally more greenish or brownish tints, and the use of dye is not disclosed.

At home, people often use coloring for decoration. It is classically used in icings for cakes and cookies to create colorful and decorative designs. Food coloring can also be used for projects like coloring Easter eggs. People also use it in craft projects, because the color usually takes well and it is nontoxic and easy to handle, which makes it great for craft projects with children.

How to Add Food Coloring to Food

Baking and cooking with food coloring adds another creative dimension to your food recipes. Working with food coloring requires you to mix primary colors according to the color wheel. However, premixed gel and paste colors are also available in the market. The best type of food to color is a white food, such as frosting or cake, which will show colors well. Choose the type of coloring that fits your style of cooking, and start coloring a single drop at a time.

Steps

1. Choose your type of food coloring. Different food coloring options have different advantages, so you may decide based on the ease of use, amount of use and richness of the color.

- Traditional liquid food coloring is available in supermarkets, craft stores and online. It comes in 4 different bottle colors: red; yellow; blue; green. It can be messy to work with, and it can permanently stain clothes in its concentrated form. You must use the basic tenets of the color wheel to mix these colors into many different shades across the color spectrum.
- Concentrated pastes are available in baking stores and craft stores. They create beautiful, rich colors; however, some people believe they produce a metallic or bitter taste, and so they can offset the taste of the food. They can be expensive, depending upon your local shopping outlet.
- Many professional bakers use soft gel pastes. They are widely available from baking outlets and online sources. These are tasteless and come in dozens of shades, so you don't have to mix them yourself. However, they can be expensive. You may need to prepare food in bulk to get good value out of them. Make sure to buy the.75 oz. (22ml) bottles rather than the 4.5 oz. (133ml) bottles.
- Research organic food coloring. If you are doing vegan baking, then most liquid and paste food coloring will be off-limits. This is because food colorings are sometimes made from insect shells and chemicals.

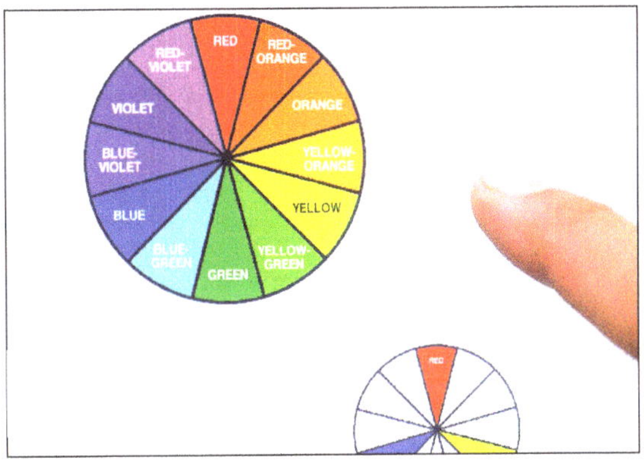

2. Get to know your color wheel. Print off a color wheel from the internet. Visiting sites that give a refresher course on color theory will help you to produce secondary and tertiary colors.

- Go to colormatters.com/color-and-design/basic-color-theory for some good images and information on the color wheel.

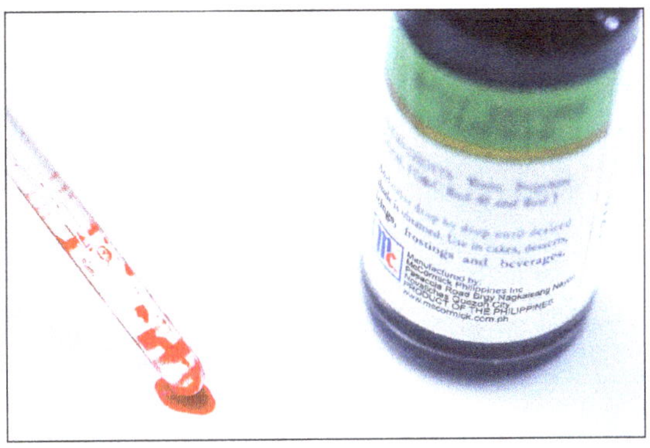

3. Start with one color, a drop at a time. You may feel as if you have to mix the colors before you add them to the food; however, this is not necessary. They will mix in if you add a drop of 1 color and drops of another color.

- Each food will respond to coloring differently. Always add the coloring 1 drop at a time.

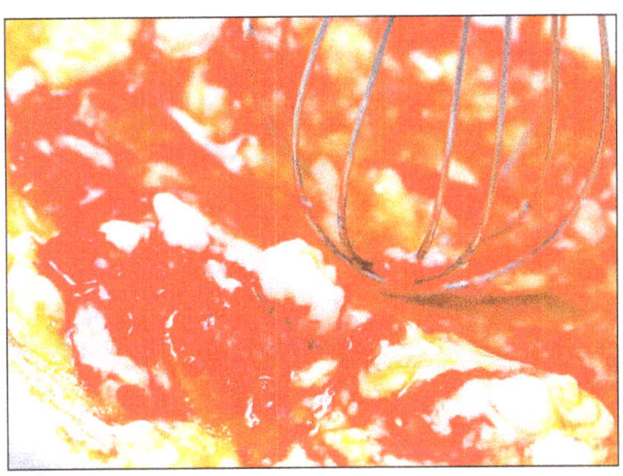

4. Add drops of colors while the food is in the mixer, if possible. If you are making baked goods, then chances are you can receive mechanical help to mix colors thoroughly.

5. Use white food coloring to make white food. Very few foods are perfectly white. If you want to have this effect, then buy white gel or paste separately.

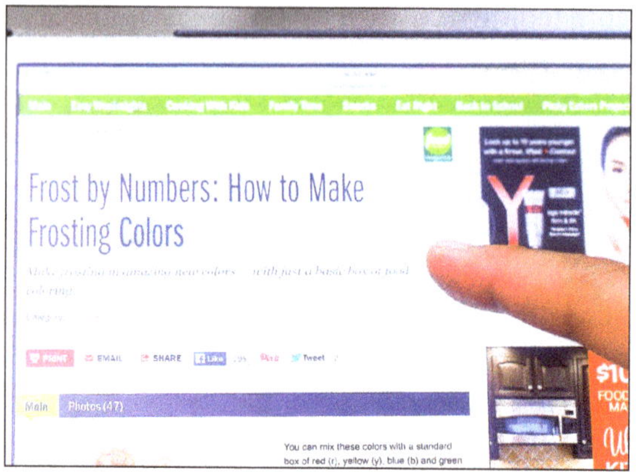

6. Consult the Food Network's "Frost by Numbers" section. This magazine and website has detailed research on how to create the most rich and unusual colors on the color wheel. Learn to make shades like pink lemonade, dinner mint and pinot noir.

- Visit Food Network's color chart to access the online version. If you want the magazine copy, you can request the May 2012 issue.

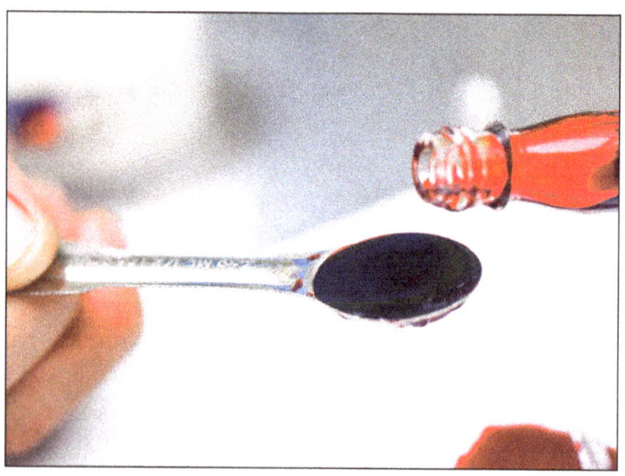

7. Use relatively large amounts of food coloring to make deep shades of red, black or brown. These shades are harder to form and may take up to 1 oz. of food coloring.

8. Color your food at least 1 hour before decorating. If you are baking or cooking the food, this may not apply. You will be able to tell if the color will fade or remain vibrant.

9. Avoid coloring foods in brown if they have a lot of acid in them. Omit the cream of tartar, vinegar or lemon juice. Using brown food coloring with this type of acidic food may result in a greenish hue.

- Acidic foods can also cause violet to look blue. Adjust accordingly.

10. Use organic food coloring, if you want to use natural methods. A number of fruits, vegetables and spices can be added to food to change the color. Make sure to use these colorings carefully, because they may change the flavor.

- Add beetroot juice to your food to give it a ruby-red color. In some foods, some of the red can bake out, but the naturally sweet juice will give a very slight sweetened boost to baked goods. Use beet juice from the can or juice your own beets.

- Juice carrots to add an orange color. These vegetables are also pleasantly sweet and good in both sweet and salty foods.
- Use saffron or turmeric very sparingly for a yellow color. These spices are strong, so research a recipe to take the potent flavor out of them before using them in sweet foods.
- For a green color, use liquid chlorophyll or spinach. A few cups of spinach gives a mild flavor but a rich color to light foods. It mixes well with fruits without adding a vegetable flavor.
- Boil red cabbage to create purple dye. Add the purple water to your recipe. Add some baking soda - a little at a time to the purple water to make a blue dye.

How to make Natural Food Coloring

Whether you're making icing for your kid's birthday cake, whipping up a vegan tofu scramble, or making your ordinary meal more holiday-appropriate, if you cook or bake then you probably have encountered a use for food coloring. Unfortunately, many food colorings currently approved for sale in grocery stores in the U.S. are believed by some to be toxic -- even potentially cancer-causing. The good news, however, is that natural food dyes are safe, easy to make, and work just as well as the artificial ones that could damage your health. You can easily color your food using everyday cooking ingredients and other food items, many of which you may already have in your pantry.

Method 1. Making Dyes from Vegetables

1. Choose vegetables with consistent coloring. Greens with dark leaves (such as spinach) and many root vegetables (such as carrots and beets) tend to work well as food dyes because they have deep, consistent, opaque coloring. Your natural food dyes will be most effective if you choose veggies that are mostly one color throughout (and which have a color that is bright or deep).

- Some vegetables may seem like good candidates for natural food dye, but if they have high water content (as does celery) they will actually produce very weak, light shades.
- Many vegetables (even those with vibrant coloring) lack the thick, concentrated juices produced by many fruits. In general, do not expect vegetable-based food dyes to be as bright or even as those made from berries. Beets (for red) and carrots (for orange) are exceptions to this rule.

2. Boil your vegetables. Some vegetables leach color into the water when boiled. The vegetables

that work best for this are those with lots of water content (and therefore juices) that are also deeply hued. Red cabbage (for purple) and beets (for red or pink) are two great examples of vegetables that can be boiled to extract their coloring.

- For a more concentrated color, only use as much water as is needed to barely cover the vegetables. The colored water becomes the dye -- the more you dilute it, the lighter the shade.
- A good rule of thumb for figuring out which vegetables produce the best colors is that those that stain your fingers when handling them will also easily dye the foods with which they come into contact.

3. Dehydrate your vegetables or herbs. Use a food dehydrator or set your oven to 150 degrees Fahrenheit and place the item(s) inside on an oven-safe tray. Cook them until they are as dry as possible (without becoming burned); this could take up to six hours.

- For large vegetables (especially sphere shaped ones), cut them into extremely thin slices before dehydrating them. This will speed up the process and dry them more consistently.
- Once dried, your vegetables can be safely stored in an airtight container for several months to a year.

4. Grind the dried veggies into a powder. Use a coffee grinder or food processor to do this most efficiently. The finer the powder, the less the dye will affect the texture of the food you wish to color.

- You can also use a mortar and pestle to grind the veggies by hand, but this will take much longer and could result in less even consistency.

- Thoroughly rinse off any utensils you use to make the powder before using them to grind up a different color of dried food. This will keep you from contaminating the color and flavor (if any) of your next veggie powder.

5. Select foods that are already in powder form. Many vegetables/herbs can be purchased in dried, powdered form, which eliminates the need to do the drying and grinding yourself. Just be sure to select those without added spices or flavoring so you don't influence the flavor of the food you wish to color.

- If you are not worried about watering down your food, you can mix the powder into a small amount of water or other liquid and then stir this into the food. Do this gradually to achieve the right color and avoid over-saturating your food.
- For yellow dye, use stale turmeric. Turmeric is often used to give vegan puddings and tofu scrambles their "eggy," yolk-like shade. Stale turmeric tends to have lost quite a bit of its natural flavor, so use this to avoid influencing the taste of your colored food.

Method 2. Making Dyes from Fruit Juice

1. Choose fruits with opaque juices. Although many fruits are very brightly colored, their juices do not always make for good food dye. Many citrus fruits, for example, have highly translucent juices (such as oranges and lemons), which do not stain other foods very effectively. Berries, on the other hand, are very effective for coloring foods.

- To figure out which fruits will work best for your purposes, squeeze or blend fruits and pour the juice into a clear glass. Hold the glass up to the light; the less light that passes through, the better the juice will work as a food dye.
- For red or pink coloring, raspberries and cherries are good options; strawberries produce a lighter, more pastel shade of pink. For blue or purple, try blackberries or blueberries.

2. Juice or blend the fruit. Unlike with boiling vegetables to leach out their color, fruit juice food coloring is made of the fruit itself. For berries, stick them in a food processor or blender in order to liquefy them for your dye. For hand fruits, you can use a juicer instead (but again, most of these types of fruits don't make good dyes).

- You can start with either fresh or frozen fruit if you are blending it, but fruits must be fresh in order to juice them effectively.
- Be sure to remove any pits, large seeds, or inedible peels before tossing fruit into the blender; these can ruin your machine and will not help your food dye production efforts.
- Add a very small amount of water to the fruit if you are using a food processor in order to allow the fruit to adequately liquefy.

3. Strain the juice. Any time you blend or juice fruits, tiny seeds, skins, or other fibers (pulp) can end up in the juice. In order to preserve the consistency of the food you wish to dye, remove these elements from the juice by passing it through a mesh sieve (with very small holes) or a cheesecloth.

- Straining is essential for maintaining consistency and smoothness in your dyes. However, if you are going for a less even look, you can choose not to strain the fruit juice (so long as there are no seeds in it).
- Very finely-blended berries will be impossible to strain completely and will most likely end up with tiny flecks of skin and fibers in them; if this is not acceptable, consider juicing or boiling instead.
- Don't use a sieve or strainer with mesh that is large enough for seeds and fibers to pass through. Test-strain a small portion of your juice to make sure your mesh is effective.

4. Reduce the juice. In some cases, the strained juice itself will be sufficient as a food dye. However, you can increase the intensity of the color by cooking off some of the water contained in the juice. Pour the juice into a small saucepan and cook it over medium heat until it becomes the consistency of a thick paste.

- This process results in a highly concentrated dye, which will have dense flavor, especially if it is made from berries. Be sure to use it sparingly to avoid flavor contamination.
- Skip this if you are going for a lighter, more pastel coloring.

Method 3. Choosing the Right Natural Coloring Sources

1. Pick compatible colors. If the food you are attempting to color is already a color other than white,

this will influence the result of your dyeing efforts. Don't expect blue frosting to turn red if you add juiced raspberries to it, for example.

- When in doubt of how something will turn out, use a small portion of dye and food to test the outcome. You can then adjust the shade of your coloring substance if necessary by mixing it with other shades.
- Avoid mixing too many different colored dyes together when trying to settle on a color. This can end up making the dye lose its vibrancy and turn out brownish.

2. Go for dye options with subtle flavors. In many cases, there are multiple options for making one color of food dye; in such cases, choose the one with the more subtle flavor. For example, both turmeric and saffron can be used for yellow coloring, but turmeric is less intense and usually makes a better choice.

- An exception to this would be if you actually want to add flavor from your dye to the food. In such cases, be sure to match the flavor types (such as sweet with sweet) to avoid creating an unappealing mixture.
- Not all dye production methods are equal in terms of intensity. Generally speaking, juicing and blending will create dyes that are both brighter and more intense tasting than the boiling or drying methods recommended for some vegetables.

3. Pay attention to consistency. To keep your food from becoming too wet while ensuring even coloring, take care to create food dyes that compliment the consistency of your food. This will take some trial and error, but common sense also goes a long way.

- For powdered coloring, stir into wet, blended foods such as cake icing or mashed potatoes. Sprinkling powder on a dry food will not spread the coloring consistently.
- For liquid coloring, use sparingly in all foods unless wetness is not a concern. Dry foods, for example, can become too soggy if too much liquid coloring is used.

4. Mask unwanted flavors when necessary. Depending on the kind of food you are coloring, you can get rid of flavor contamination from your food dye by adding additional flavors. For instance, a hint of flavor from beet juice coloring in cake icing can be easily covered up with a drop or two of vanilla or peppermint extract.

- This method doesn't work quite as well for savory foods colored with sweet dyes. If you color cauliflower red using raspberry puree, for example, even lots of salt and butter might not be able to mask the sweetness of the dye.
- Many extracts and other substances that can be useful for masking flavors, such as truffle oil, are quite expensive (and potent), so use them sparingly.

How to make Black Food Coloring

Black food coloring can be hard to find in stores. Luckily, you can prepare your own black food coloring right at home.

Method 1. Mixing Food Coloring

1. Buy red, blue, and green food coloring. You can mix these colors to make a dark grey, which is about as close as you can get without buying black food coloring.

- If you're making icing or frosting, use gel or paste food coloring. Liquid food coloring is less intense and can make the icing runny.

2. Mix in cocoa powder (for white frosting only). The end result is always better when you start with a dark recipe. If you're using white frosting, you can fix this by mixing in cocoa powder, one spoonful at a time.

- Black cocoa powder gives the best result, but ordinary cocoa powder should work fine for this method.
- If you skip this step, you'll need to use much more food coloring, which can affect the taste and texture.

3. Add equal amounts of red, blue and green to your recipe. Start with just a couple drops of each

color, stirring them in well. Repeat until the mixture darkens to a deep grey, always adding equal amounts of the color.

- You can use yellow instead of green, but the lighter color makes it more difficult to achieve black.

4. Adjust the color. If you see hints of other colors in the grey, make these adjustments:

- If it looks green, add more red.
- If it looks purple, add more green.
- Make all adjustments one drop at a time, stirring thoroughly between each drop.

5. Wait for the final color. Most food coloring will deepen in buttercream and fade slightly in royal and boiled icing. If you are making the latter type of icing, consider adding the color a half hour before serving to minimize the fading.

- In some regions, chemicals in the water can change the color. Buttercream usually has a more reliable color when made with milk instead.
- Keep the recipe away from direct light and heat, which can cause fading.

Method 2. Using Natural Ingredients

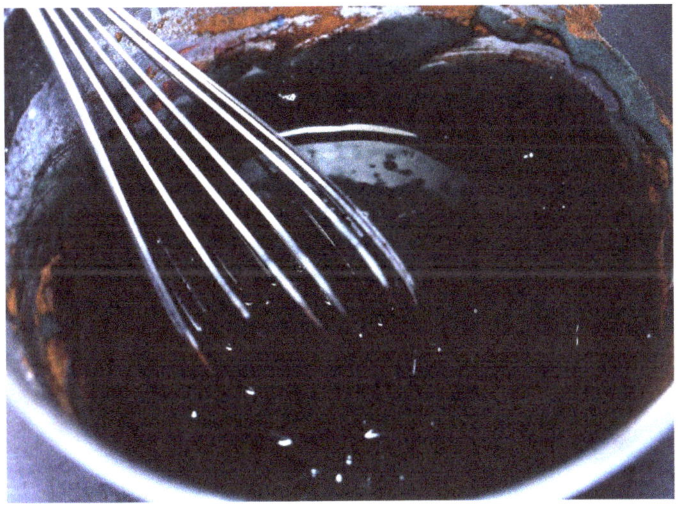

1. Mix black cocoa powder into cake batter. This special type of cocoa powder, labeled "black" or "ultra Dutch processed," has a darker color and more mild taste than regular cocoa. This will make a deep black cake with a chocolatey flavor. When replacing ordinary cocoa powder, make the following changes to the recipe:

- Add a little extra fat (butter or oil).
- Use 1 tsp (5mL) baking powder instead of each ¼ tsp (1.25mL) baking soda.

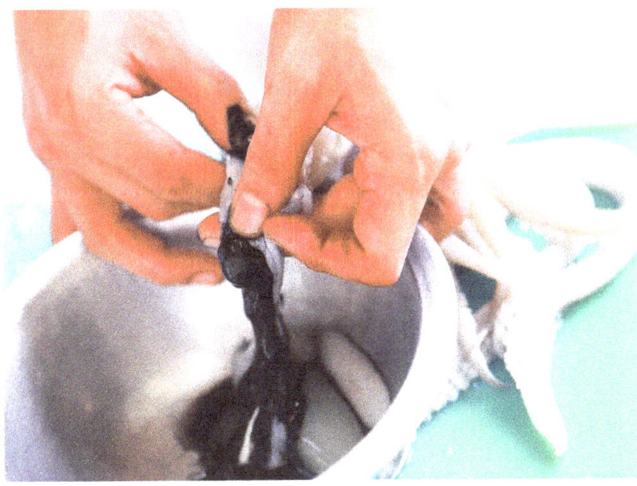

2. Add squid ink to savory dishes. This has a briny taste and is not suitable for sweets or desserts. This is mostly used to color pasta, rice, or savory sauces. For the most intense color, mix it into homemade pasta dough (replacing the salt and a little of the liquid ingredients). For a faster but less consistent method, add the ink to the water while the pasta or rice is boiling. Stir ink into the sauce as well for a more dramatic look.

- Fishmongers sometimes sell this, but you may need to ask at several stores.
- Add squid ink in small quantities. It is quite salty and can taste of iodine in large quantities.

How to make Brown Food Coloring

If you don't have brown food coloring on hand or don't want to go with the store-bought variety, there are ways to mix it up yourself at home. Read on to learn how to blend colors for that perfect brown and how to use common edibles like cocoa and coffee to make a natural brown with a little flavor kick.

Method 1. Mixing Food Coloring

1. Find two colored dyes that are opposites on the color wheel. Red and green, orange and blue, or yellow and purple can all be combined to make brown.

- You can use liquid or gel food dye. Gel food dye is usually preferred because you don't need very much to get vivid colors.

2. Mix the color in a 1:1 ratio. So for every 1 drop of red, add 1 drop of green.

- Mix your colors in a glass or stainless steel bowl. The dye can stain plastic bowls.

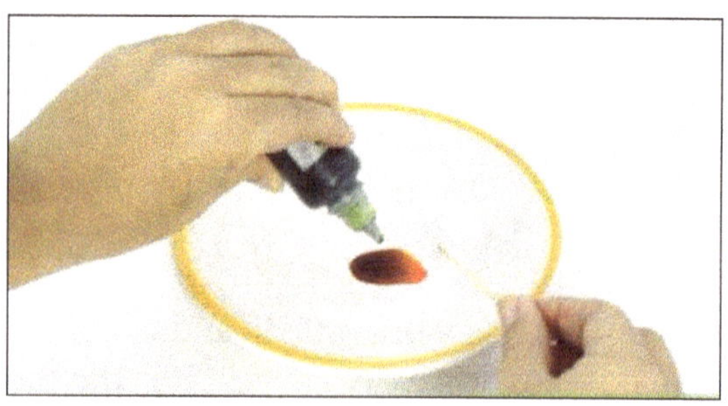

3. Play with the addition of colors until you reach the desired shade of brown.

- Mix red and yellow in a 1:10 ratio (1 drop red to 10 drops yellow) for tan.
- Add 1 drop blue for a darker brown.
- Add red or yellow if the dye looks gray or black.

- Add orange and green to lighten a red and green mix and make tan.
- Add 1 drop of black dye for a very dark brown.

4. Remember that brown dye will darken as it dries. You may think you have the perfect shade of espresso brown and then find the next day that is has deepened to almost black. Mix your dye lighter than you think you want so it has room to develop.

- If you are dying frosting, consider making it 1-2 days ahead of time so you know the true shade you are working with.
- If your frosting still isn't dark enough, add a drop of black dye or more of your brown.
- If your frosting is too dark, add a little of the white frosting to dilute the color.

5. Add your dye a little at a time. It often takes a lot of brown dye to make a rich color, but you risk ruining your frosting, playdough, etc. by drowning it in liquid.

Method 2. Making Natural Food Coloring

1. Choose your colorant. The colorant is what will provide the brown coloring for the dye. You can use ground espresso, instant coffee, cocoa powder, cinnamon, or black tea achieve brown.

- Remember that you may need a lot of dye to get a very dark brown, which means you will also get a stronger taste of the colorant. Cinnamon is great for light brown, but the taste might be overwhelming if you make a dark brown.
- Used in small amounts, you should not taste the colorant. It's ok to use a dill seed-based colorant for your buttercream if you use it very sparingly.
- You may not be able to get the same vivid colors with natural dye as you would with store-bought colors. Petroleum-based colors need only a few drops to add color, whereas you might need a lot of natural dye to add color. Most recipes will be ruined by the addition of too much coloring--they will taste weird or become too watery to use.

2. Steep colorants in water to make liquid dye. For every 1 cup of water, add 1 cup coffee, 1/2 cup cocoa, 10 tea bags, or 4 tablespoons of spice.

- Bring the water and colorant mixture to a boil.
- Boil gently or simmer between 15-30 minutes.
- Allow mixture to cool to room temperature.
- Pour mixture through a fine mesh strainer to catch any unabsorbed granules. Store in a glass or stainless steel bowl, as the dye will not stain these materials.

3. Grind colorants in a coffee or spice grinder to make a powder dye. The key to a powder dye is that it must be extremely fine. If any of your colorants are coarse (like crystals of instant coffee), grind them further until they are a powdery consistency.

- Powder dyes are more concentrated, so you may be able to get a darker color using less of the powder dye. Keep in mind that too much can change the flavor of whatever you are dyeing.

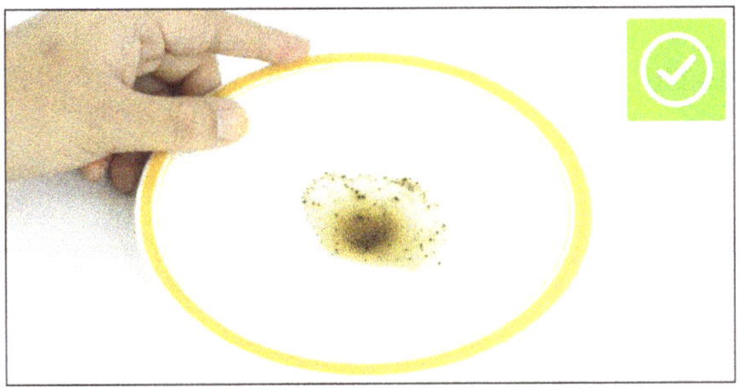

4. Add your dye a little at a time. Remember that too much liquid dye can ruin the texture and structure of whatever you are dyeing, especially frosting. Too much liquid or powder dye can affect the flavor, so be sure to taste as you go.

- Make peace with the idea that you may not get the dark, vivid color you want from natural dye. Flavor and consistency is ultimately more important than getting that perfect shade.

How to make Purple Food Coloring from Blackberries

Making purple food dye from blackberries is a cheap and simple option. This topic outlines a simple way to do this.

Steps

1. Obtain 2 cups worth of fresh blackberries and wash them.

2. Put the blackberries in the blender and turn on until the berries are completely pureed.

Alternatively, if you do not have a blender, thoroughly mash the berries in a bowl with a mixer or wooden spoon, or force them through a strainer.

3. Strain the mixture through a strainer to remove any solids.

4. Put your coloring into cake, or anything else you want it for.

How to make Blue Food Coloring

Blue food coloring can be used to turn many different foods and beverages, like frostings, water, and cocktails, blue. While you can buy blue food coloring in any grocery store, you can also experiment with your own homemade version using red cabbage. The blue you'll achieve with homemade food coloring will be more muted and subtle than what you can buy in the store, but if you're looking for a fun DIY project, making homemade blue food coloring is a great choice.

Part 1. Extracting the Color from Red Cabbage

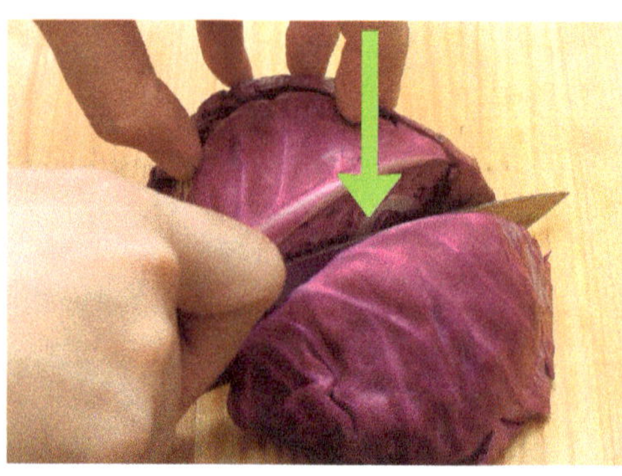

1. Chop the cabbage. Place an entire head of red cabbage onto a cutting board. Cut off the stem and

discard it. Hold the cabbage steady with one hand, and with a large chef's knife, roughly chop the cabbage into small chunks. The cutting doesn't have to be precise, but smaller pieces are better because you'll get more color from the cabbage.

- Red cabbage contains anthocyanin, a purplish pigment that will dissolve in water. You can collect purple cabbage water to turn it into blue food coloring

2. Rinse the cabbage. Transfer the cabbage pieces to a large bowl. Cover the cabbage with water and use your hand to stir the cabbage around in the water. Pour the cabbage into a colander to drain out the excess water.

3. Boil the cabbage in water. Transfer the cabbage to a large saucepan. Cover the cabbage with enough water to cover it completely, about 8 cups (1.9 L). Put on the lid and bring the water to a boil over medium-high heat. When the water comes to a boil, reduce the heat to medium, and continue simmering for 20 minutes.

- As the cabbage boils, the anthocyanin will leach into the water, turning it purple.

4. Strain the cabbage. Remove the pot from the heat. Insert a colander into a large glass or metal

mixing bowl. Pour the cabbage and the purple water into the colander. The colander will catch the cabbage, and the purple water will drain into the bowl below.

- The cabbage is cooked and ready to eat. Set it aside to cool, and then use it in soups, stews, stir fries, and other dishes.

Part 2. Turning Cabbage Water into Blue Food Coloring

1. Boil down the purple water. Pour the collected purple water into a medium saucepan. Bring the water to a boil over high heat, and then reduce the heat to medium-high. Continue simmering the water with the lid off for about 75 minutes to reduce the purple water. Cook the mixture until most of the water has evaporated and there's only ½ cup (118 ml) of liquid left.

- As the liquid cooks, the water will evaporate and the liquid will reduce, leaving behind a more concentrated and more vibrant purple water.

2. Let the mixture cool. When the water has reduced, remove the saucepan from the heat. Set the pan aside to cool to room temperature. This could take about 30 minutes. When the purple concentrate has cooled, pour it into a clean mason jar or other clear container.

- Use a clear jar so you can see the color change as you add the baking soda.

3. Add some baking soda. Measure ⅛ teaspoon (0.6 g) of baking soda and pour it into the jar with the purple concentrate. Stir the mixture to dissolve the baking soda in the water, and allow the bubbles to subside if necessary.

- The purple pigment in the cabbage reacts differently with acids and bases. In an acidic environment, the anthocyanin will become redder. When you mix the anthocyanin with an alkaline ingredient like baking soda, it will become bluer.

4. Test the color. Measure ¼ cup (59 ml) of a white liquid, such as milk, soy milk, or coconut milk, into a clear glass. You can also test the color on a small amount of frosting before dyeing the entire batch. Add a few drops of the food coloring and stir to incorporate the dye into the liquid. The color will be pale, but it will give you a good idea of the blue shade you've achieved.

- The testing step is particularly important if you're going to use the dye to color frosting or something similar, because the pH of the frosting will alter the color of the dye.

5. Add more baking soda if necessary. Continue adding baking soda in ⅛-teaspoon (0.6-g) increments and testing the color until you achieve the desired shade of blue. Don't add more than ½ teaspoon (2.4 g) of baking soda. Otherwise, the food coloring will taste salty and metallic.

- The baking soda won't make the color darker, but it will make the dye bluer.

Part 3. Using Homemade Blue Food Coloring

1. Make blue ice cream. Homemade ice cream is a fun and tasty project, and the best part is that you can completely customize the ice cream according to your taste. Transfer the ice cream base to the ice cream maker, turn on the machine, and add the blue food coloring a few drops at a time until you achieve the desired intensity.

- You can also add blue food coloring to store-bought white ice cream. Portion out a serving of ice cream and set it aside to soften for a few minutes. Before serving, add several drops of blue food coloring and stir to turn the ice cream blue.

2. Make blue frosting. Blue frosting is great for cupcakes and cakes, Halloween decorations, Easter treats, and much more. Make a batch of vanilla buttercream or your favorite white frosting. Before icing your treats, add several drops of food coloring and stir to combine. Add more drops if necessary until you achieve the desired color.

- Note that frosting colored with homemade blue will have a much more pastel hue than with store-bought food coloring.

3. Color Easter eggs. Dyeing eggs is a fun tradition to celebrate Easter. Hard boil the eggs first to help prevent cracking, and when the eggs are dry, dip them into a jar of the blue food coloring for several minutes. Alternatively, you can paint the eggs with the dye if you want to make patterned eggs.

4. Use the color to make fun cocktails. There are many different cocktails and drinks that are blue,

including the blue lagoon and the blue Hawaiian. To turn your drinks blue, add 1 to 2 tablespoons (15 to 30 ml) of blue food coloring to any clear or light-colored liquids, such as water, vodka, or even milk.

- Many blue drinks use an alcoholic liqueur called blue curaçao for the blue coloring, but you can use your food coloring instead to make non-alcoholic versions.

5. Avoid using the color for foods that will be heated. When you bake foods like cakes and cookies, chemical reactions occur when the different ingredients interact and heat up. These reactions can change the color of your dye from blue to something unexpected.

- You can experiment with your blue food coloring and baked goods, but be warned that the dye may react unpredictably, and could turn a brown or unpleasant color.

6. Store leftovers in the refrigerator. If you have any leftover blue food coloring, transfer it to an air-tight container and keep it chilled in the refrigerator. The food coloring will last for at least 2 weeks.

7. Finished.

How to use Powdered Food Coloring

Food coloring powders are a great way to alter the appearance of food without affecting the taste. They work well with foods that are not meant to be mixed with water. While dying food is easy, achieving a desired color with the powders may take some trial and error. The powders are highly concentrated; use them sparingly. Use them in the most natural light possible to achieve the truest color representation. After the color is mixed with food such as icing, let the mixture sit for a few minutes prior to use for the color to darken fully.

Coloring Icing

Open the shades on windows to let natural light into the workspace. Fluorescent lighting alters the shade of colors; working in as much natural daylight as possible will make achieving a desired color shade easier.

Spoon the icing into a bowl. Sprinkle a little food coloring powder on top of the icing and stir.

Allow the colored icing to sit for five minutes. The color will fully develop in this time. If you need a darker color, sprinkle a little more powder into the bowl and stir. Allow the color to develop again before deciding if the color is right or not. Repeat if desired.

Painting Food

Set up in a naturally lit workspace.

Pour some lemon extract or drinking alcohol into a bowl. Sprinkle some food coloring powder in with the liquid and stir.

Allow the colored liquid to sit for five minutes. The color will fully develop in this time. Sprinkle a little more powder into the bowl and stir if you want a darker shade. Allow the color to develop again before deciding if the color is right or not. Repeat if desired.

Dip a food-grade brush in the colored liquid to apply to food.

Dusting

Add the food coloring powder to sugar or other dry ingredients.

Stir the powder and other dry ingredients together.

Use this mixture for dusting food as you would use granulated sugar, coconut flakes and other dry products.

Tips

- Powdered food colors do not mix with water.

How to Mix Liquid Food Coloring

Food coloring that you find at grocery stores typically comes in four colors: red, blue, yellow and green. Some stores carry larger lines of specialized shades, but most likely you'll have to mix colors to achieve the soft hues preferred for baking. These steps are focused on coloring icing or frosting; however, basic color-mixing theory carries over to many other projects in which you are coloring a white food. And speaking of icing, before you start mixing in the colors, you might want to pull out a few tablespoons to use later as a lightener in the event your color mix is darker than you had hoped.

Step 1

Download a basic color wheel or color chart. Find one that shows both secondary and tertiary colors. You don't need this to follow the steps below, but it's a handy reference guide that will help you learn basic color theory — eventually, it will become second nature, and you won't need to consult a chart.

Step 2

Obtain at minimum the following primary colors: blue, red and yellow. You cannot create these colors by mixing others, but with them you can create any of the other shades in the color wheel.

Step 3

Mix equal parts of any two primary colors to produce the secondary colors of green, orange and violet. One drop of red mixed with one drop of yellow makes orange; one drop of yellow mixed with one drop of blue makes green; and one drop of blue mixed with one drop of red makes violet.

Step 4

Create secondary colors by adding only one drop each of the two primary colors. Mix thoroughly. Continue to add more drops in equal amounts to achieve the color you want.

Step 5

Use unequal amounts of primary colors to achieve tertiary shades shown in the outer ring of the color wheel. For instance, if your goal is to achieve a light orange, you would use more yellow than red, like one drop of red against two drops of yellow. After stirring, if you want to make the color darker, add more red; if you want it lighter, add more yellow or stir in more white icing.

Step 6

Create some eye-catching tertiary colors using the following formulas:

Aquamarine = 2 drops of blue + 1 drop of green

Chartreuse = 1 drop of green + 1 drop of yellow Magenta = 2 drops of red + 1 drop of blue Gold = 3 drops of yellow + 1 drop of red

Step 7

Try mixing colors separately for better results when creating pastels. For instance, to create lavender icing, start by adding a tiny swirl of red to create pink in one portion of white icing. Then add a drop of blue to separate portion of white icing. Combine small amounts of the blue icing with the pink icing to create lavender.

Step 8

Fine-tune the color of the icing by making minor corrections. Make it darker by adding more of the darker primary color. Make it lighter by adding in some of the extra white icing.

How to make Vegetable Dyes for Easter Egg Crafts

While there are commercially available dye sets for Easter egg dyeing blown out eggs, it's also fun making your own vegetable dyes for eggs that you can eat at the dinner table. For kids, this provides a great lesson in making do from the garden or kitchen resources. Here is how to make vegetable dyes for coloring real egg Easter eggs that can be enjoyed at the dinner table.

Steps

1. Prepare the eggs first. Put the eggs into a saucepan and cover with water. Bring to the boil and

boil for 10 minutes. You want them to be absolutely hard boiled. Remove and allow to cool. Once cooled, remove their shells.

2. Make yellow dyes. Use turmeric powder. Simply add turmeric powder to hot water in a ceramic or glass bowl (plastic might stain). Add the eggs to the bowl and leave to soak up the dye for one to two hours.

3. Make orange or orange-brown dye. Use onions. Keep the skins off 10 onions for the week leading up to dyeing the eggs. Boil the skins in two cups of water until the water turns brown. Drain the water into a bowl and discard the onion skins. Add the eggs to the brown water and soak for one to hours to allow the dye to penetrate the shell.

4. Make pink or red eggs. Use beetroot. Pour the juice from a can of beetroot into a suitable bowl. Add the eggs to soak for one to hours.

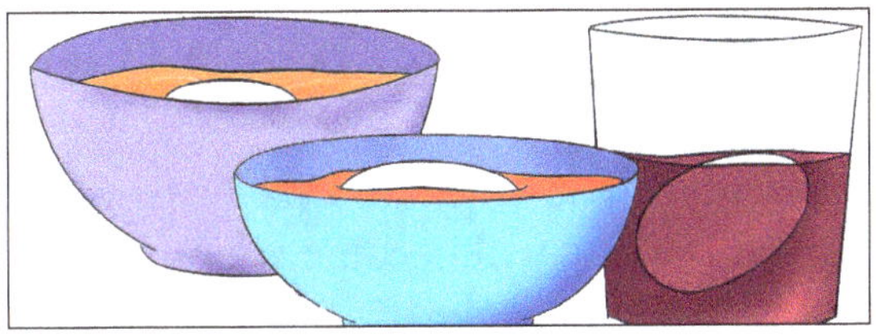

5. In each case, keep an eye on the soaking process to determine the color that pleases you most.

6. Dry the eggs once you're happy with the color. Leave them to sit on a baking tray or cookie sheet.

7. Serve. Once the eggs are dried, make a centrepiece from them in a glass bowl or dish. Place in the centre of the table along with a serving implement and encourage diners to help themselves. The flavor of the food used to dye the egg will be an additional taste delight, along with the pretty colors.

8. Experiment with other vegetable or food dyes from the kitchen. There are lots of other possible colors that you can create from items in the kitchen, from a variety of vegetable or fruit juices from the can to the cooking water left after steaming greens. Just be sure to choose flavors that you think people will enjoy.

How is Frosting Made with Natural Food Colors

We're sharing four all-natural ways to add color and flavor to your next batch of buttercream. Light, airy, and smooth, this hued frosting will become your go-to recipe. A mix of vibrant fruits and veggies are all you need for natural food coloring.

Ingredients

- 8 egg whites
- 1 1/2 cups sugar
- 1/2 teaspoon cream of tartar
- 1/4 teaspoon salt
- 6 sticks (1 1/2 pounds) unsalted butter, cut into 1-tablespoon-size pieces, room temperature
- 1 vanilla bean, split and beans scraped
- 2 teaspoons vanilla
- 1 1/2 cups fresh strawberries, halved

Directions

1. In a very large heatproof bowl continually whisk egg whites, sugar, cream of tartar, and salt over a pan of simmering water until sugar dissolves and mixture reaches 160 degrees F.

2. Remove from water. Beat with a mixer on high speed until stiff peaks form. Add butter, a small piece at a time, until mixture thickens and becomes smooth and spreadable (mixture will look thin and a little curdled, but will thicken as butter is added). Stir in vanilla bean seeds and vanilla.

3. In a food processor or blender process or blend strawberries until smooth. Press through a fine mesh strainer (you should have 3/4 cup puree). Beat or whisk strained strawberry mixture into the frosting.

Naturally Colored Buttercream Variations

- Berry Buttercream: Prepare as above, except substitute 3/4 cup blackberries and 3/4 cup blueberries for the strawberries.
- Mango-Carrot Buttercream: Prepare as above, except substitute 4 mangoes, seeded, peeled, and cut up, for the strawberries and stir in 1/3 cup carrot juice with the sieved mango puree.
- Peach-Spinach Buttercream: Prepare as above, except substitute one 16-ounce package frozen peach slices, thawed, for the strawberries. Place 3 cups fresh spinach and 1/3 cup water in a blender. Cover and blend until smooth. Press through a fine-mesh sieve. Stir into frosting with the sieved peach puree.

Chapter 5

Food Preservatives and Preservation Methods

Food preservation is aimed at preventing the growth of microorganisms, slowing down the oxidation of fats causing rancidity and inhibiting the visual deterioration of food. Several methods of preservation of food exist such as drying, dehydrating, smoking, freezing and vacuum packing. This chapter discusses these preservation methods and their effective uses.

Food Preservatives

All food products except for the one growing in your kitchen garden has food preservatives in them. Every manufacturer adds food preservative to the food during processing. The purpose is generally to avoid spoilage during the transportation time.

Food is so important for the survival, so food preservation is one of the oldest technologies used by human beings to avoid its spoilage. Different ways and means have been found and improved for the purpose. Boiling, freezing & refrigeration, pasteurizing, dehydrating, pickling are the traditional few. Sugar, mineral salt and salt are also often used as preservatives food. Nuclear radiation is also being used now as food preservatives. Modified packaging techniques like vacuum packing and hypobaric packing also work as food preservatives.

Food Preservation is basically done for three reasons:

- To preserve the natural characteristics of food.
- To preserve the appearance of food.
- To increase the shelf value of food for storage.

Natural Food Preservatives

In the category of natural food preservatives comes the salt, sugar, alcohol, vinegar etc. These are the traditional preservatives in food that are also used at home while making pickles, jams and juices etc. Also the freezing, boiling, smoking, salting are considered to be the natural ways of preserving food. Coffee powder and soup are dehydrated and freeze-dried for preservation. In this topic the citrus food preservatives like citrus acid and ascorbic acid work on enzymes and disrupt their metabolism leading to the preservation.

Sugar and salt are the earliest natural food preservatives that very efficiently drops the growth of bacteria in food. To preserve meat and fish, salt is still used as a natural food preservative.

Chemical Food Preservative

Chemical food preservatives are also being used for quite some time now. They seem to be the best and the most effective for a longer shelf life and are generally fool proof for the preservation purpose. Examples of chemical food preservatives are:

- Benzoates (such as sodium benzoate, benzoic acid)
- Nitrites (such as sodium nitrite)
- Sulphites (such as sulphur dioxide)
- Sorbates (such as sodium sorbate, potassium sorbate

Antioxidants are also the chemical food preservatives that act as free radical scavengers. In this category of preservatives in food comes the vitamin C, BHA (butylated hydroxyanisole), baterial growth inhibitors like sodium nitrite, sulfur dioxide and benzoic acid.

Then there is ethanol that is a one of the chemical preservatives in food, wine and food stored in brandy. Unlike natural food preservatives some of the chemical food preservatives are harmful. Sulfur dioxide and nitrites are the examples. Sulfur dioxide causes irritation in bronchial tubes and nitrites are carcinogenic.

Artificial Preservatives

Artificial preservatives are the chemical substances that stops of delayed the growth of bacteria, spoilage and its discoloration. These artificial preservatives can be added to the food or sprayed on the food.

Types of Artificial Preservatives Food

- Antimicrobial agents
- Antioxidants
- Chelating agent

In antimicrobial comes the Benzoates, Sodium benzoate, Sorbates and Nitrites.

Antioxidants include the Sulfites, Vitamin E, Vitamin C and Butylated hydroxytoluene (BHT)

Chelating agent has the Disodium ethylenediaminetetraacetic acid (EDTA), Polyphosphates and Citric acid

Harmful Food Preservatives

Although preservatives food additives are used to keep the food fresh and to stop the bacterial growth. But still there are certain preservatives in food that are harmful if taken in more than the prescribed limits.

Certain Harmful Food Preservatives are Benzoates

This group of chemical food preservative has been banned in Russia because of its role in triggering allergies, asthma and skin rashes. It is also considered to cause the brain damage. This food preservative is used in fruit juices, tea, coffee etc.

Butylates

This chemical food preservative is expected to cause high blood pressure and cholestrol level. This can affect the kidney and live function. It is found in butter, vegetable oils and margarine.

BHA (Butylated Hydroxyanisole)

BHA is expected to cause the live diseases and cancer. This food preservative is used to preserve the fresh pork and pork sausages, potato chips, instant teas, cake mixes and many more.

Caramel

Caramel is the coloring agent that causes the vitamin B6 deficiencies, genetic effects and cancer. It is found in candies, bread, brown colored food and frozen pizza.

In addition to this there are many other harmful food preservatives. These are Bromates, Caffeine, Carrageenan, Chlorines, Coal Tar AZO Dies, Gallates, Glutamates, Mono- and Di-glycerides, Nitrates/Nitrites, Saccharin, Sodium Erythrobate, Sulphites and Tannin.

Preservatives Food Additives

All of these chemicals act as either antimicrobials or antioxidants or both. They either inhibit the activity of or kill the bacteria, molds, insects and other microorganisms. Antimicrobials, prevent the growth of molds, yeasts and bacteria and antioxidants keep foods from becoming rancid or developing black spots. They suppress the reaction when foods comes in contact with oxygen, heat, and some metals. They also prevent the loss of some essential amino acids some vitamins.

Some Common Preservatives and their Primary Activity

Chemical Affected	Organism(s)	Action	Use in Foods
Sulfites	Insects & Microorganisms	Antioxidant	Dried Fruits, Wine, Juice
Sodium Nitrite	Clostridia	Antimicrobial	Cured Meats
Propionic Acid	Molds	Antimicrobial	Bread, Cakes, Cheeses
Sorbic Acid	Molds	Antimicrobial	Cheeses, Cakes, Salad Dressing
Benzoic Acid	Yeasts & Molds	Antimicrobial	Soft Drinks, Ketchup, Salad Dressings

There are other antioxidants like Sodium Erythorbate, Erythorbic Acid, Sodium Diacetate, Sodium Succinate, Grape Seed Extract, Pine Bark Extract, Apple Extract Tea Proplyphenols, Succinic Acid and Ascorbic Acid and food preservatives like Parabens and Sodium Dehydro Acetate used frequently for preservation.

How to Add Preservatives

Preservatives come naturally and chemically synthesized. They work to stop the growth of harmful microbes, allowing food to remain safe to eat longer. People have been practicing preserving techniques for many centuries. The French submerge slow-cooked goose in its own rendered fat, cooling and storing to preserve. This is called a confit (pronounced con-fee) and is recognized as a delicacy, although it started as a traditional means of preserving meat in fat.

Preserve fruits with sugar by making jam or candy. Sugar kills bacteria, or slows its growth, by drawing water away. Bacteria needs moisture to survive. Fruit naturally contains sugar in the form of fructose and glucose. Many recipes do not require you to add sugar in order to preserve fruits. Europeans have done this for hundreds of years, making pommé (apple jelly), poiré (pear jelly) and raisiné (grape jelly). Bring any kind of very ripe fruit to a boil, reduce the heat and then cook until the jam does not run off a spoon. Stir frequently. Pour jam into canning jars and seal with lids and rings. Turn upside down to sterilize any air remaining in the jar. The jam will keep for two years.

Preserve fruit or vegetables by pickling with vinegar. In order to safely preserve food, vinegar must have an acidity level of 5 percent. For cucumbers, wash and cut firm fruit. Bring one packet of quick process pickling mix and 4 cups of vinegar to a near boil in a nonmetal or coated pan. Metal reacts with vinegar and can produce a cloudy pickle solution. Pack raw cucumbers into canning jars and pour simmering pickle liquid over them, keeping 1/4-inch head space. Cover each jar with a lid and seal with a ring. Process with a water-bath canner, according to your canner's directions.

Cure meat with salt, sodium nitrate and sodium nitrite. People have used salt to preserve meat and fish for centuries. Salt prevents bacteria by inhibiting growth. Commercially-produced sodium nitrite and sodium nitrate preserve meat by slowing the development of botulism and preserving color. Using too much nitrate or nitrite can be toxic so stick with a commercially-produced, premixed cure. Morton Salt makes home-curing kits you can find in the meat or spice section at the grocery store.

Use only safe, FDA-approved chemical preservatives like sodium benzoate. For use as an antimicrobial and yeast-fighting agent, sodium benzoate must be used with naturally acidic food or food that has been acidified. Sodium benzoate comes in granular or powder form and should be stored in an airtight container until you are ready to use. Add small amounts to dressings, beverages, pie fillings, etc. Only 0.1 percent is allowed in food products by law. You can find sodium benzoate at kitchen supply stores or wholesalers.

How to Vacuum Pack Food

Vacuum packing food sucks all the oxygen out of the packaging. This makes the food last 3 to 5 times longer and preserves it appearance because microorganisms, such as bacteria, grow at a much slower rate inside of a vacuum seal. Also, freezer burn is virtually eliminated because the food is not in contact with cold air. For someone who plans to regularly vacuum pack food, purchasing a machine that automates the process may be worth the additional cost. Here is how to vacuum pack food using either an automatic machine or manual pump.

Method 1. Vacuum Pack Food with an Automatic Machine

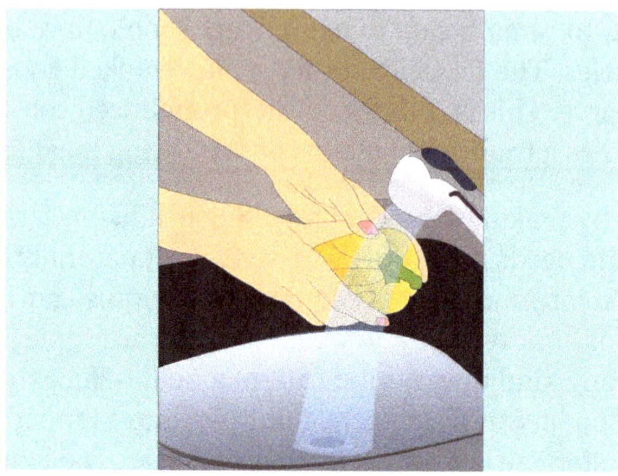

1. Clean and prepare the food that you wish to vacuum pack.

- Scrub or peel fruits and vegetables so dirt is not trapped inside the vacuum seal.
- Trim any skin and bone that you don't need from meat.

2. Place the food inside the plastic bag. Most machines require using a specific brand of bag.

3. Put the open edge of the bag into the sealing machine.

4. Press the button, if necessary, to start the vacuum and sealing process. Many vacuum pack food machines have an automatic sensor that alerts the machine when a bag has been placed inside and automatically starts the sealing process.

5. Watch as the vacuum pack food machine withdraws the air and the plastic bag shrinks.

6. Wait for the machine to stop its operation, which signaling that the plastic bag has properly been sealed.

7. Remove the vacuum packed food and store it in the pantry, fridge or freezer.

Method 2. Vacuum Pack Food with a Manual Pump

1. Clean and prepare the food that you wish to vacuum pack.

2. Insert the food into the plastic bag or container. (Many manual vacuum sealing systems use containers rather than bags.)

3. Place the lid on the container or seal the bag.

4. Put the nozzle end of the pump into the provided hole on the plastic bag or container lid.

5. Press down on the pump repeatedly until all the oxygen is sucked from the bag or container.

6. Remove the nozzle from the pumping hole. (Many of the manual systems have a one way nozzle so that air won't escape once the nozzle has been removed.)

7. Store the food in the pantry, fridge or freezer.

How to Preserve Food by Drying

Food drying offers an alternative method of food preservation than canning or freezing food. The process involves extracting moisture from specific foods to prevent the formation of bacterial growth and other organisms that cause rot or decay. Drying works best when it occurs outdoors in an oven or in a food dehydrator. Vegetables, fruits, herbs, meats and fish can withstand the food-drying process. Certain preparation methods apply before drying foods.

Method 1. Prepare for Drying Food

1. Wash your hands thoroughly with soap and water and clean any utensils you plan to use.

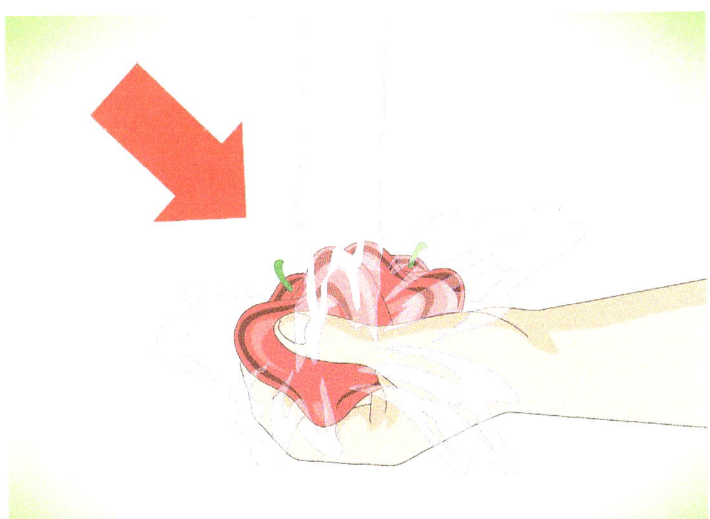

2. Clean fresh vegetables, such as green beans, corn and peppers.

    - Trim away any bruises with a knife. Cut vegetables into smaller pieces, at least 1/2-inch (3.8 cm) thick, for easier drying.

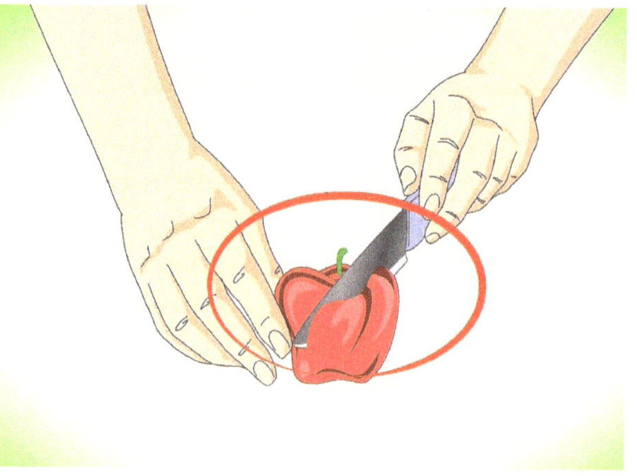

- Place 1 quart (0.94 liters) of vegetables into a 36-inch (91 cm) cheesecloth.

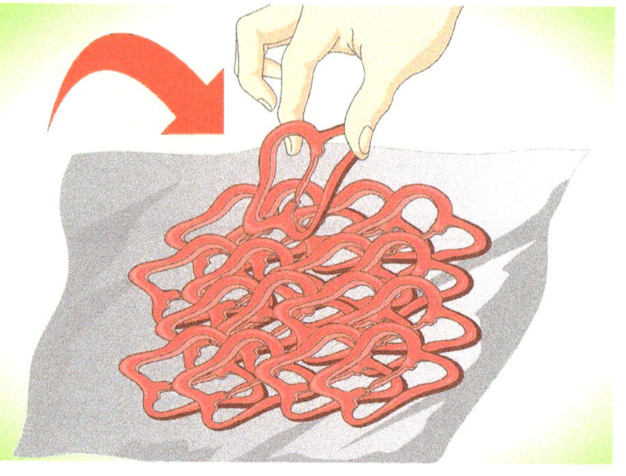

- Blanch the cheesecloth bag of vegetables on the stove in a half-filled pot of boiling water. You may also add 1/4 teaspoon (1.2 ml)of citric acid to the water. Allow up to 6 minutes for blanching, then immediately place the bag in cold water for the same amount of time.

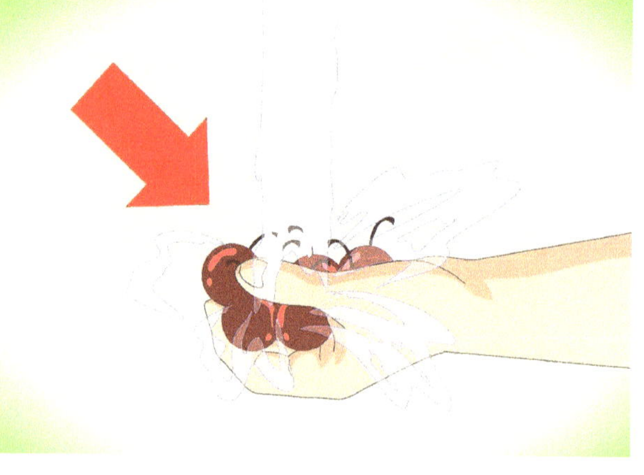

3. Rinse and slice fruits, like apples, cherries and peaches.

   - Cut the fruit into 1/2-inch (1.27 cm) pieces or in halves to ensure even drying. Fill a large pot with at least 1 quart (0.94 liters) of water.

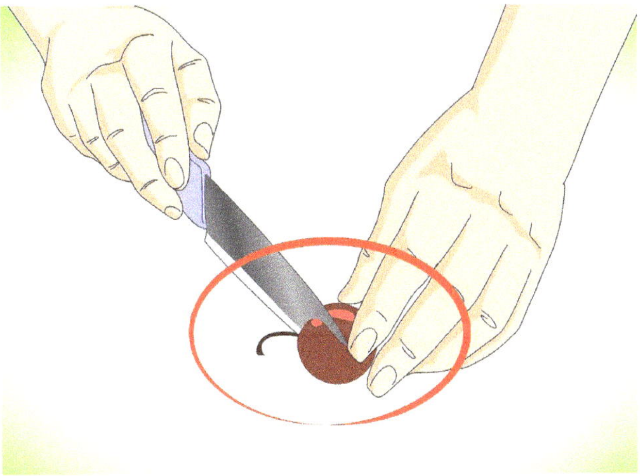

- Add up to 1.5 teaspoons(7.3 ml) of food-grade quality sodium sulfate or sodium bisulfate to the water. Soak for approximately 5 to 15 minutes, depending on how you sliced the fruit. Rinse the fruit pieces with cold water before placing them on a rack to dry.

4. Gather herbs during the morning once the dew evaporates.

- Collect the herbs right after the morning dew to minimize the risk of wilting.

- Rinse the herbs in cool water and shake off any excess.

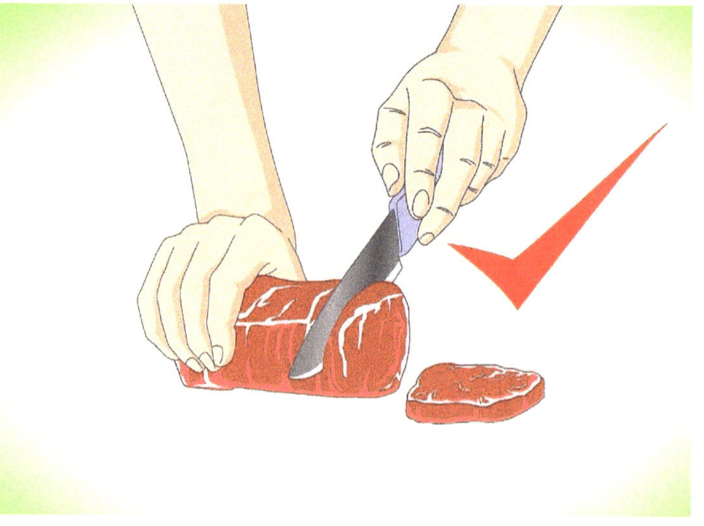

5. Clean and cut lean meats and fish.

- Slice lean cuts of beef or salmon into 1/4- to 1/2-inch thick (1.27 cm) slices. Cut off all fat because it spoils the meat during the drying process.
- Freeze any meat or fish you plan to dry for around 30 days. The freezing temperature should be at 0 degrees Fahrenheit (approximately 17 degrees Celsius) or lower.
- Thaw the meat and fish in the refrigerator. Add salt, seasonings and any marinades of your choice before the drying process. Remember to wash your hands again after handling vegetables, fruit, herbs and meat.

Method 2. Choose your Drying Method

1. Arrange slices of vegetables, fruits or meats in single layers on drying trays. Place the trays inside of an oven or a food dehydrator.

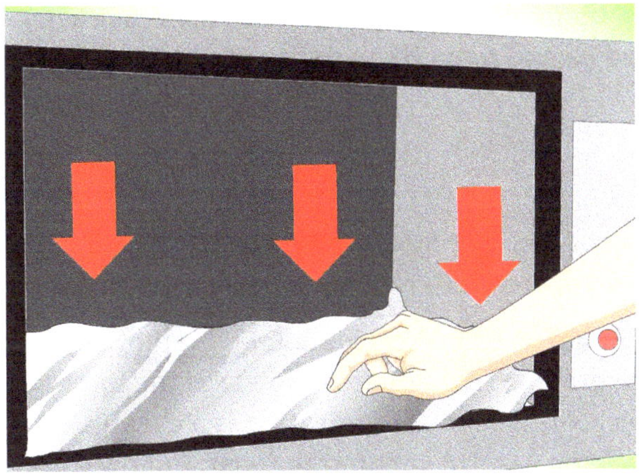

2. Line the bottom of the drying appliance with aluminum foil if you marinated the meat. This will catch the drippings.

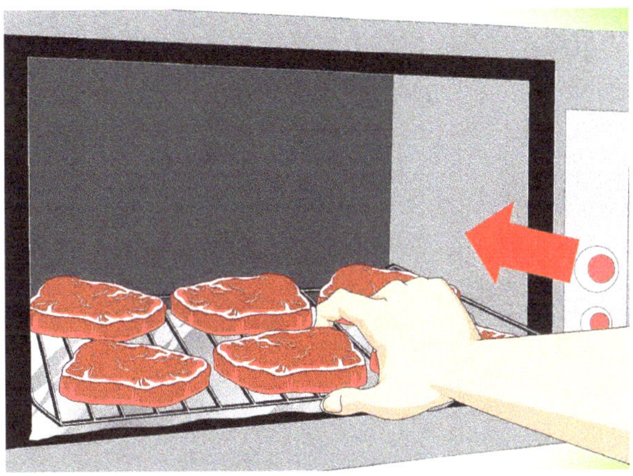

3. Place the drying trays in the appropriate appliance.

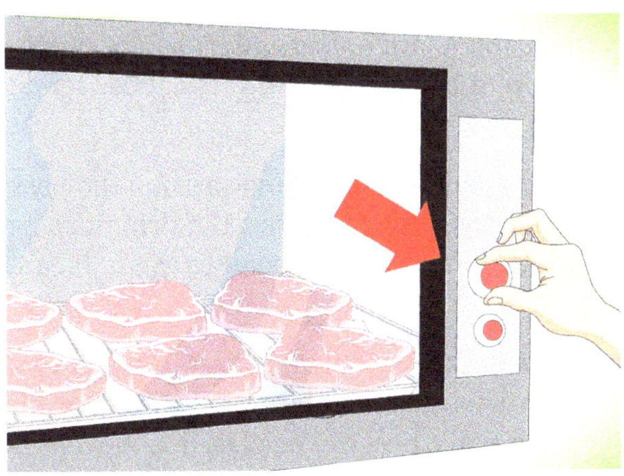

4. Start the drying process.

- If you use an oven, then leave the door slightly open and keep a fan nearby for ventilation.
- Close the door of the food dehydrator since it already comes with a ventilation fan.

- Air-dry herbs or put them in a food dehydrator. Tie herbs such as basil, sage, rosemary and thyme in small bundles and hang them outdoors. Place the herbs in a food dehydrator for faster drying.

How to Preserve Food by Dehydrating

Dehydrating, or drying, is a means of preserving food by removing its water content. Almost any food with water can be dehydrated. This not only increases the food's shelf life, it can also prevent decay and the growth of microorganisms. Dehydration is a cheap alternative to canning foods and is a sure way to make your supply last throughout the year. Follow these steps for dehydrating your food.

Part 1. Choosing a Food Dehydrator

1. Invest in a vertical dehydrator if you are only drying a few or one type of food. Vertical dehydrators have heat travel from top to bottom or vice versa. They tend to be smaller but are also cheaper.

- Vertical dehydrators with a fan on the bottom provide the best heat distribution and are more efficient because hot air rises. However, the drippings from fruits, vegetables, and meats can drip into the fan and become difficult to clean. It can also cause the dehydrator to malfunction or fail.
- Having a fan on top eliminates this problem but tends to dry foods on the top more quickly than those on the bottom. This may not always be a problem, since certain foods have different drying times. For example, beef can be placed in the top tray (less water content) and apples on the bottom (more water content). The main disadvantage here is that the foods dried together vertically can adopt each other's tastes.
- American Harvest/Nesco Food Dehydrators offer a variety of top-fan/bottom-fan dehydrators at a wide range of affordable prices.

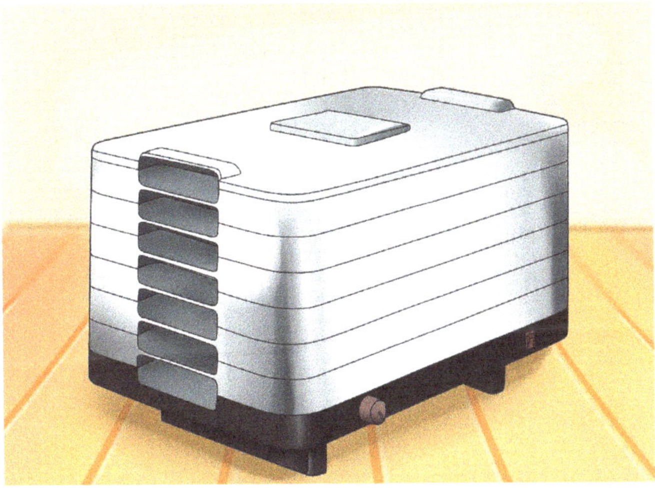

2. Use a horizontal dehydrator if you are drying large quantities or many different foods. This model of dehydrator tends to be large, allowing multiple foods to be dried at once. It also provides the most even heat distribution.

- Horizontal dehydrators have the fan or the principal drying element in the back of the unit. Since the air doesn't pass directly from one tray to the next, this minimizes the imparting of flavors between foods. This means your beef jerky won't taste like your apple chips, and vice versa.
- The main disadvantage is that horizontal dehydrators are more expensive.
- Excalibur Dehydrators are well built and a very popular brand among drying enthusiasts.

3. Choose a food dehydrator with a fan if you are making bark or fruit leather. Some dehydrators use other heating mechanisms to dry food, which often takes longer and yields poor results in fruits.

- Using a dehydrator without a fan can cause banana slices to dry unevenly, sometimes making it feel moist and other times making it feel like poker chips. As a result, drying fruit becomes an unpredictable and inefficient process.

4. Buy a dehydrator with adjustable temperature settings. Different foods require different temperatures; one temperature fits all is not a good strategy for drying food.

- Look for a range of 95–155 °F (35–68 °C). Meat is typically dried in the 145–155 °F (63–68 °C) range, while fruits and vegetables hover around the 125–135 °F (52–57 °C) range.
- Temperature is very important when it comes to dehydrating food. Too low a temperature can cause food to spoil while too high a temperature can cause the surface of the food to harden, preventing moisture from escaping.
- Know that lower priced dehydrators, while frugal, also may not offer different temperature settings.

5. Purchase the right trays and accessories depending on the foods you're drying. Tray size isn't the only feature to consider when drying foods of different sizes, textures, and water content.

- Tray size isn't that great of a factor unless you are seeking to dry large quantities of food. Otherwise, your food should dry at a predictable and consistent rate, provided you have purchased a quality dehydrator.
- Mesh sheets are necessary for drying small vegetables like peas and corn. They are pliable, making it easier to pop off fruits like bananas, which tend to stick to plastic when dried. These sheets are also necessary for certain tray structures intended for meats. Meat trays will have gaps where slices of fruit will fall through without mesh sheets.

- For blended foods like mashed potatoes, tomato sauce, and fruit purees, buy non-stick sheets or fruit leather trays. Non-stick sheets are reusable and work better than parchment paper. Never use wax paper as it will melt in the dehydrator.
- Stacked trays are a feature of certain dehydrators and make it difficult to check the food you are drying. Sliding trays allow you to slide a tray out rather than removing it completely, making the monitoring process much easier.

Part 2. Dehydrating Meat

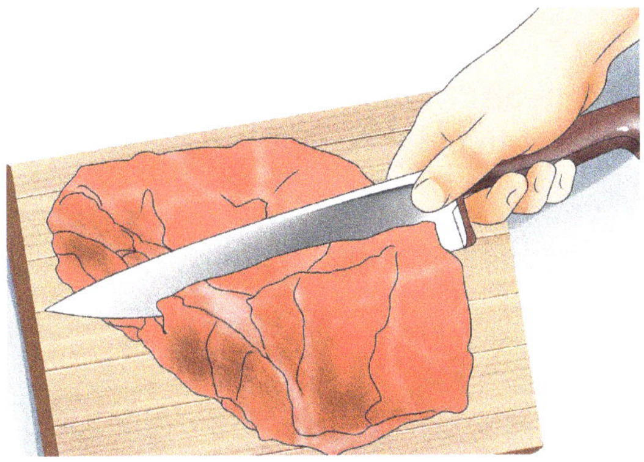

1. Cut your meat into smaller slices. Be sure your slices are uniformly cut so that dehydration occurs consistently throughout your meat.

- Slice ham into 1-inch wide strips. They should look like slightly thicker slices of bacon.
- Cut beef into long, 1/4-inch wide strips if you are making beef jerky.
- Pull chicken apart into small pieces. It should look very similar to pulled pork.
- If you plan on consuming meats after they have been dried, be sure that your ham and chicken are precooked. Raw, dehydrated beef is okay to consume as it will turn into beef jerky. Eating raw dehydrated pork can cause an infection called trichinosis, which occurs when raw or under-cooked pork is consumed. Similarly, eating raw chicken can give you salmonella food poisoning.

2. Place your strips of meat onto a tray and place it inside the dehydrator. Position slices into neat rows so that they do not overlap or cover one another. Spread the pulled chicken into an even layer so that there are no large clumps.

3. Dry the meat at 145–155 °F (63–68 °C) for about 6 hours. These times and temperatures may vary slightly for different meats but should yield an overall similar result.

- If making beef jerky, monitor your slices to make sure they are pliable but not brittle. This means they should bend without actually breaking.

4. Dab ham and beef slices with paper towels frequently throughout the drying process. The moisture that comes to the surface will mostly be oil and fat from the meat.

- Oils and fats don't evaporate as easily as smaller molecules like water. Therefore, you need to wipe it away in order for drying to be successful.
- You don't have to wipe the pulled chicken bits as chicken is leaner and contains less fat.

5. Take the meat out of the dehydrator once it is completely dry. Use your fingers to test the meat and see if there is still moisture on the surface.

- Drying requires a lot of monitoring and is not a precise procedure like baking, for instance. Don't be afraid of opening up the dehydrator to check the meat every few hours to observe its progress.

6. Store your dehydrated meats in airtight plastic bags. Remember, air also contains moisture, and moisture is the enemy of dried foods.

- If you're storing the meat shorter than a month, store it in a dry, dark place at room temperature. Your kitchen cupboard should be ideal for dried meats. Don't worry about spoilage; removing the water content prevents meats from spoiling.
- For long term storage, place meat in the freezer or refrigerator.

7. Check your dried meats every couple of weeks or so. Even though water has been pulled from the meat, air may still find a way of getting in. Since air tends to carry bacteria and other organisms, sealed meats can still become moldy or spoiled.

- When storing dried foods, contamination from insects can occur. Don't worry, however, since this is not common among properly sealed meats. It is most likely a result of meats already containing the eggs prior to drying.
- To reduce insect contamination, pasteurize your meat after drying it. You can either store the meat in the refrigerator for 48 hours or place it in the oven at 175 °F (79 °C) for 15-30 minutes.
- Dried foods can be stored up to one year. Vacuum-packing and refrigerating can double or triple this shelf-life.

Part 3. Dehydrating Fruits and Vegetables

1. Wash and dry your fruits and vegetables. Even though most of the germs will be killed during the drying process, you want to reduce the amount of bacteria as much as you can beforehand.

2. Blanch all vegetables except onions, peppers, and mushrooms. Blanching will help preserve the taste and texture of crunchier vegetables.

3. Slice your fruits and vegetables into even slices. Be sure to peel and core fruits like peaches, apricots, apples, pineapples, and pears before dehydrating.

- For corn, cut the actual corn off of the cob instead drying the entire vegetable.
- Remove the seeds from peppers after you have sliced them.
- You can leave mushrooms whole if you wish.

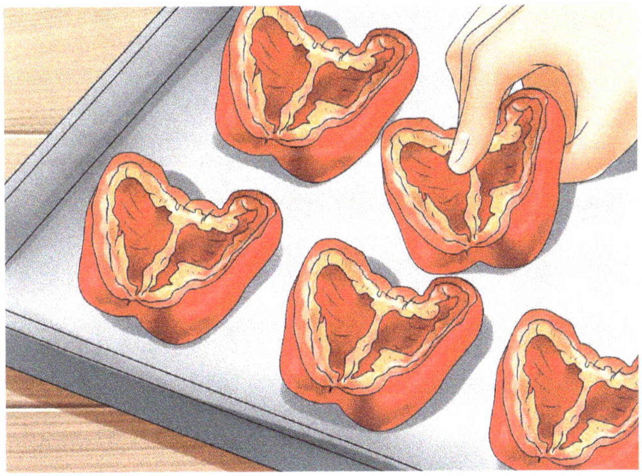

4. Lay your sliced fruits or vegetables onto the tray in a single, even layer. If you are drying many different fruits/vegetables at a time, designate one tray for each kind.

- Try to limit how many fruits and vegetables you dry at a time. Even if you are using a horizontal dehydrator, having too many items can throw off drying times.

5. Dry the larger fruits or vegetables at 130–135 °F (54–57 °C) for 6-12 hours. For smaller vegetables like corn, broccoli, mushrooms, and peas, drying for 3-10 hours will be sufficient.

- These times vary from plant to plant and depends largely on the water content of the food. Most fruits dry at the same temperature and for the same amount of time, but some vegetables have drastically different drying times.
- The most different drying times belong to corn, broccoli, mushrooms, and peas. Since these vegetables are small and contain less water, they tend to dry in half the time of other vegetables.

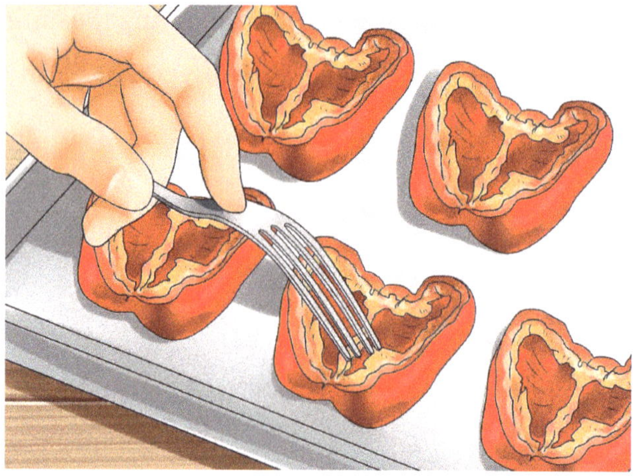

6. Check for specific textures as you dry your fruits or vegetables. Dryness textures will vary from plant to plant, so be sure to read these specific textures for each fruit and vegetable.

- Green beans, carrots, corn, peas, mushrooms, and zucchini should all feel brittle.
- Beets, peppers, blueberries, cherries, pears, and pineapples should all feel leathery.
- Onions, potatoes, and tomatoes should feel crisp. Bananas and strawberries should feel *almost* crisp.
- Apples, apricots, peaches, and strawberries should all feel pliable.
- Broccoli and cauliflower will simply feel dry and hard.

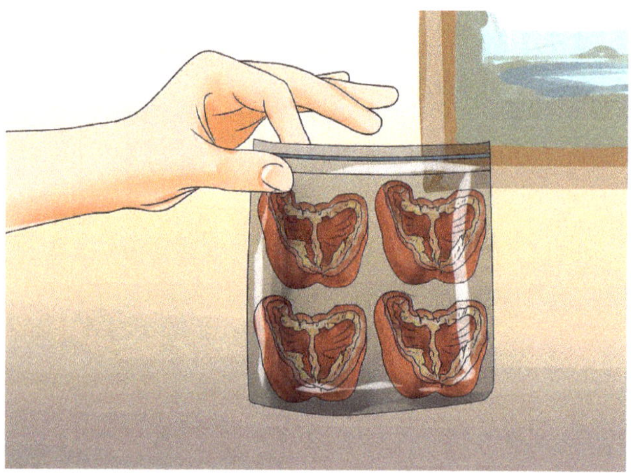

7. Store dried fruits similarly to how you would store meats. For periods shorter than one month, store in vacuum sealed containers in dry, dark areas. For periods longer than that, store in refrigerators and freezers.

- Keep fruits and vegetables high in Vitamin A away from direct light. Vitamin A is sensitive to light and is retained during the drying process. Fruits containing vitamin A - like carrots, bell peppers, and mangoes - can degrade in direct sunlight.
- For the best quality, fruits and vegetables should be replaced every year.

How to Preserve Food by Freezing

Freezing can be a convenient way to store extra food and cooked meals to be enjoyed at a later date. However, the way you package the food up makes all the difference when it comes to preserving flavor, texture, color, and nutrient content. Package the food tightly in flat, easily stored shapes. Label all of your packages with the food name and packaging date so you can keep track of when certain things expire.

Method 1. Packing Food for Freezing

1. Choose freezer-friendly foods. Meat, fish, vegetables, fruits, bread, raw pastry, grated cheese, butter, margarine, milk, casseroles, soups, stews, chili, and meatloaf all freeze well. Avoid freezing raw eggs, cooked grains and pastas, most fried food, mayo, sour cream, fruit jelly, meringue, gelatin, and milk sauces.

2. Season lightly before freezing dishes. Freezing causes many seasonings and spices to change flavor. Spices like cloves and pepper, and foods including green pepper, garlic, imitation vanilla, and some herbs, will have a strong, bitter taste after being frozen. Onion and paprika may also shift in flavor, and celery seasonings get stronger. Curry will also taste off and slightly musty.

- If you're cooking large portions to freeze for later, season lightly. You can add seasoning to your liking once the food has been defrosted and reheated.

3. Let hot food cool at room temperature before refrigerating it. Hot and recently cooked foods need to be chilled in the refrigerator before freezing. However, they need to cool down at room temperature for about 20 to 30 minutes before putting them in the fridge. You can just leave the dish out on the counter to cool down.

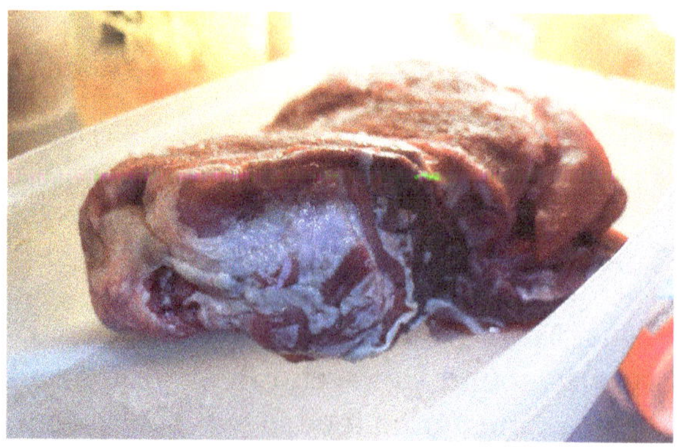

4. Chill room temperature food in the fridge before freezing it. After cooling for 20 to 30 minutes, transfer the food to a wide, shallow dish. Place the shallow dish in the fridge, uncovered. Chill the food until it reaches refrigerator temperature.

- If you put hot or warm food in the freezer, it will raise the temperature inside, causing frozen foods already in the freezer to partially thaw and then refreeze. This can negatively impact the quality, taste, and texture.

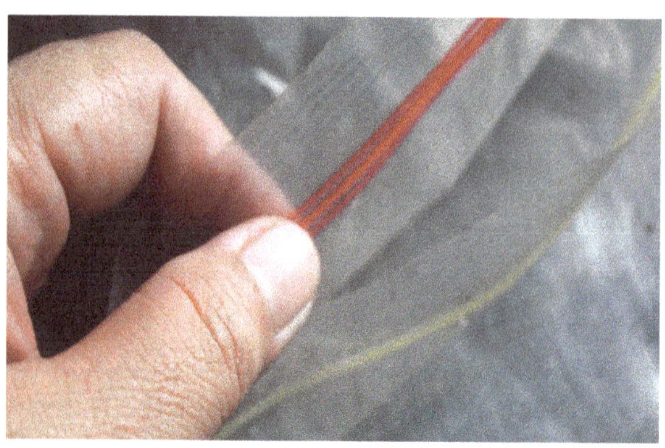

5. Use high quality packaging and freezer bags. No matter what size you choose, make sure the

bags are flexible and labeled as "freezer bags," not "storage" or "sandwich" bags. Use moisture resistant wrapping materials like plastic freezer wrap, freezer paper, and heavy-weight aluminum foil. Rigid containers made of plastic or glass also work great because they're easy to stack in the fridge.

- Make sure the containers have airtight lids.
- Choose materials that are easy to write on or otherwise label.

6. Pack foods in single-meal portions. Not only do foods freeze faster when packaged in smaller portions, but the food is easier to defrost and use later. Because of bacteria growth, it's not safe to defrost a large portion of food, use some of it, and refreeze the leftovers. The only exception is if you cook the food before you refreeze it. The heat will kill the bacteria.

- In general, avoid freezing more than one quart of food in a single package.
- The faster a dish freezes, the better the quality will be once it's thawed.

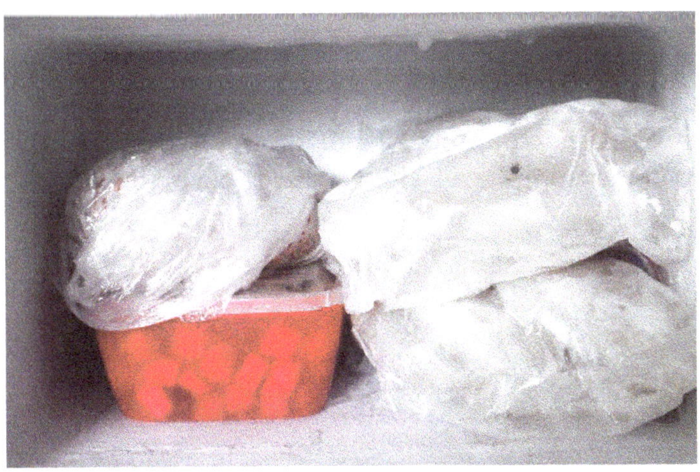

7. Arrange the food so that the package has a flat, square shape. Thin, flat packages of food are easier to store because you can stack them in your freezer. Flat, rectangular packages also freeze more quickly and defrost evenly. Round packages defrost unevenly because they take longer to defrost in the center.

8. Seal food tightly in the bag, container, or packaging. Freezer bags have 2 types of closures – the fold-over types that are secured with twist ties, and the zipper or pressed seal closures. All of them work fine as long as you remove as much air as possible from the bag before you seal it up. Make sure the seal is tightly closed.

- Use freezer tape on packages wrapped in butcher paper.
- Check containers for a firm seal before putting the food in the freezer.

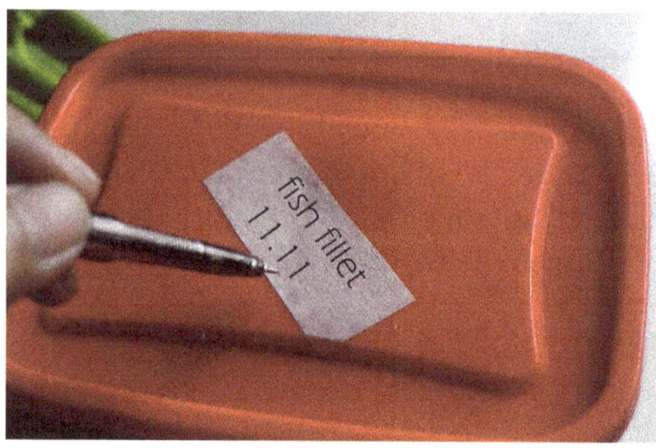

9. Label the package clearly with the dish name and packaging date. To avoid having a freezer full of mystery foods, label every package with the dish name and the packaging date. You may want to write the number of servings and any other helpful information on the outside of the package, as well. Use a permanent marker so the label will remain legible in the months to come.

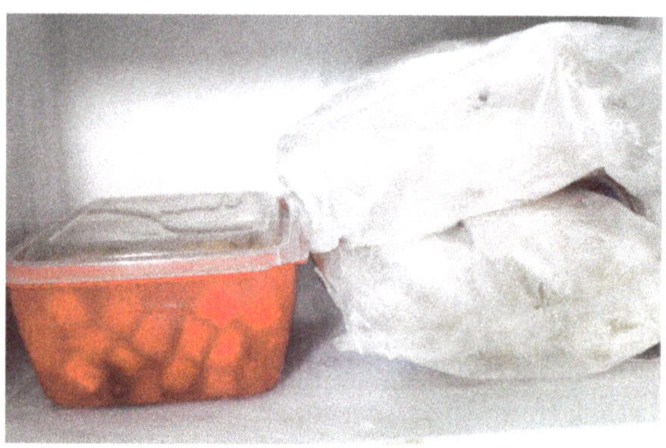

10. Put the packages in the freezer with plenty of space around them. Avoid packing frozen food

tightly in your freezer. When you first put the food in the freezer, arrange the packages in a single layer with a little room around each one. That way, the air circulates and the food freezes faster. Once fully frozen, you can stack them to make room in the freezer for other items.

Method 2. Freezing Chicken, Meat and Fish

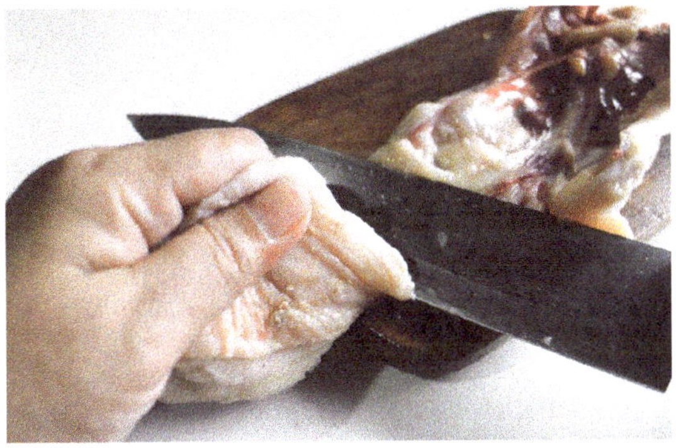

1. Trim the fat and bones from all meat before freezing. This will release excess fluids and gases that are present in the meat, helping to retain the freshness and quality during freezing. Store meats from large cuts, like steaks and venison, in the freezer for up to 12 months.

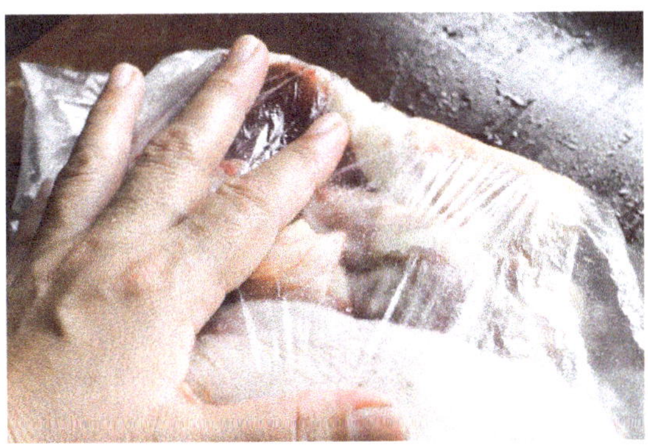

2. Freeze chicken for up to 9 months. Transfer chicken to freezer bags and remove all the air from each bag before sealing it. You may want to wrap the bags in aluminum foil, plastic wrap, or freezer paper before putting them in the freezer. This will protect against freezer burn and add an extra layer of protection against contamination.

3. Store meats for the appropriate length of time. The shelf life varies from meat to meat, mostly

depending on how much fluid is present in the meat. Store cured ham and bacon in the freezer for up to 1 month, cooked meats for up to 2 months, and ground meat in the freezer for up to 3 months. Keep sliced and shaved meat and hotdogs in the freezer for up to 2 weeks.

4. Freeze most fish for up to 3 months. Fatty fish, like salmon, go rancid the fastest. Toss those out after 2 to 3 months. Lean and cooked fish will last from 4 to 6 months. Store oysters for 2 to 3 months. Shrimp and scallops will be fine for 3 to 6 months. Store other shellfish between 2 and 3 months. Lobster, however, can be stored for up to 12 months.

5. Double wrap fish to control the smell. When freezing fish, package them tightly and then wrap them again. This will prevent that fishy smell from filling your freezer and tainting the other foods in there.

Method 3. Freezing Fruit and Veggies

1. Stick to using freezer-friendly veggies. Many vegetables freeze well, but there are some that do

not. In general, cabbage, celery, cress, cucumbers, endive, lettuce, parsley, radishes, potatoes, and veggies with a high water content will not freeze well. They will lose their flavor and often become limp once thawed.

2. Blanch vegetables before freezing them. Most vegetables retain their color, flavor, and texture best when blanched first. The time will vary for different vegetables, but in general, boil the vegetables for 3-5 minutes, then plunge them into an ice bath. Dry the vegetables with a paper towel and package them up.

- Peppers and onions are the exceptions; do not blanch these vegetables.
- Always wash the vegetables before blanching.

3. Throw vegetables out after 8 to 12 months in the freezer. Most frozen veggies will retain freshness and flavor for up to one year. After that, toss them out. The sooner you consume the vegetables, the better they will taste.

- Toss out frozen tomatoes after 2 months.

4. Wash and trim your fruit before freezing. This will help retain the freshness of your fruit and prevent it from turning brown in the freezer. Rinse the fruit under fresh, running water, then slice each fruit into pieces. Pat the pieces dry with a paper towel, then package them up in well-sealed containers or bags.

5. Store most fruits in the freezer for 10 to 12 months. Most fruits can be frozen for up to a year. After that, toss them out. Citrus fruits have a shorter shelf life – they will retain their flavor and appearance for 4 and 6 months.

- Fruits with high water content, like watermelon, do not freeze well.
- Toss out bananas after 2 months.

How to Preserve Food by Smoking

Smoking is the process of flavoring, cooking, or preserving food by exposing it to the smoke from burning or smoldering plant materials like wood and that too hardwood. Smoking of food is done at the commercial as well as the home level. The meats are hung or placed on racks in an enclosed area so that the smoke does not escape and instead penetrates foods. Commercial smoking is sometime carried out with steam pipes that supplement the wood fire.

Since smoke is antimicrobial and antioxidant it is used to preserve foods. Smoking is probably the oldest food preservation method and could have developed shortly after fire was discovered. Due to the use of chemical preservatives the use of smoking as a preservative had declined but once again it is gaining popularity as it is considered natural. The most common smoked foods are meat and fish though cheeses, nuts and vegetables are also smoked as also beverages like tea and whisky. In Scandinavia smoking the fish is popular while in Europe and in the USA it is ham.

Hot Smoking

Hot smoking foods are exposed to smoke in a controlled environment. Hot smoked foods can be reheated or cooked but they are safe to eat even without further cooking. Hot smoking is carried out in temperatures that are within the range of 52°C to 80°C. This temperature range is enough to cook foods fully and yet retain the moisture and flavor. If the temperature goes beyond 85°C foods can shrink since both moisture and fat can be dried away. Smoke roasting and smoke baking are two processes that combine hot smoking with roasting and baking. This process can be carried out in closed wood fire pit.

Cold Smoking

Cold smoking is used more as a flavor enhancer. It is used specially for cold smoking pork chops, beef steaks, chicken breasts, salmon and scallops. The temperatures used for cold smoking are between 20 to 30°C when done in a smokehouse. This temperature is enough to retain moisture in foods and does not cause the meats to harden but provides a surface which smoke can easily penetrate for the smoked flavor. However, cold smoking alone is not sufficient for preserving food as cold smoking does not cook foods. Since meats will not cook fully so before cold smoking, meats need to be cured. Curing meats removes moisture required by bacteria to grow. The smoke also works as a barrier to prevent bacteria growth.

Health Concerns

Smoked foods must be eaten sparingly as smoked, cured and salted meats are seen as stomach cancer risk. Smoking can create compounds like polycyclic aromatic hydrocarbons, or PAHs which are considered carcinogens.

Chapter 6

Food Presentation

Food presentation is the art of decorating, processing, arranging and modifying food for enhancing its aesthetic appeal. Foods may be decorated with icing, drizzled with sauces, topped with edible ornamentations, sprinkled with toppings, powders and seeds, or garnished. The various techniques of food presentation along with the aspects of plating of food, using Nasturtiums in food, etc. have been elaborated in this chapter.

Food presentation is the art of modifying, processing, arranging, or decorating food to enhance its aesthetic appeal.

The visual presentation of foods is often considered by chefs at many different stages of food preparation, from the manner of tying or sewing meats, to the type of cut used in chopping and slicing meats or vegetables, to the style of mold used in a poured dish. The food itself may be decorated as in elaborately iced cakes, topped with ornamental sometimes sculptural consumables, drizzled with sauces, sprinkled with seeds, powders, or other toppings, or it may be accompanied by edible or inedible garnishes.

Historically, the presentation of food has been used as a show of wealth and power. Such displays often emphasize the complexity of a dish's composition as opposed to its flavors. For instance, ancient sources recall the hosts of Roman banquets adding precious metals and minerals to food in order to enhance its aesthetic appeal. Additionally, Medieval aristocrats hosted feasts involving sculptural dishes and shows of live animals. These banquets existed to show the culture and affluence of its host, and were therefore tied to social class. Contemporary food aesthetics reflect the autonomy of the chef, such as in nouvelle cuisine and Japanese bento boxes. Dishes often involve both simplistic and complex designs. Some schools of thought, like French nouvelle cuisine, emphasize minimalism while others create complicated compositions based on modern aesthetic principles. Overall, the presentation of food reflects societal trends and beliefs.

Palatschinke, a thin filled pancake, here presented on a plate with chocolate sauce, whipped cream, and powdered sugar

Elaborately decorated tea party birthday cake

Plating

The arrangement and overall styling of food upon bringing it to the plate is termed plating. Some common styles of plating include a 'classic' arrangement of the main item in the front of the plate with vegetables or starches in the back, a 'stacked' arrangement of the various items, or the main item leaning or 'shingled' upon a vegetable bed or side item. Item location on the plate is often referenced as for the face of a clock, with six o'clock the position closest to the diner. A basic rule of thumb upon plating, and even in some cases prepping, is to make sure you have the 5 components to a dish; protein, traditionally at a 6 o'clock position, vegetable, at a 2 o'clock position, starch at an 11 o'clock position, sauce and garnish.

How to Present Food on a Plate

1. Choose the Perfect Plate

Selecting the right plate for your meal is key to attractive food presentation. Here are some things to keep in mind:

- Choose the right plate. One way to conceptualize plating is to think of yourself as an artist, the plate as your canvas, and the food as your medium.

- Choose the right size plate Choose your plate wisely by making sure it's big enough to allow your food to stand out, but small enough that your portions don't look too small.
- Choose a complementary plate color. The color of your plate is also significant. White plates are popular because they create high contrast and provide a neutral background for your colorful creations. Utilize white space by thinking of the rim as your frame, and consider using the rule of thirds to highlight your plate's focal point(s). When applied to cooking, the rule of thirds prescribes placing the focal point of your dish to either the left or right side of the plate, rather than the center.

2. Placing your Ingredients

Here are a few of the most important aspects to consider as you build your dish:

- Plate with a clock in mind. As you begin plating your ingredients, picture the face of a clock. From the diner's point of view, your protein should be between 3 and 9, your starch or carbohydrate from 9 and 12, and your vegetable from 12 and 3.

- Use moist ingredients as your base. Another rule of thumb is to plate moist or runny ingredients first, as they tend to move during delivery if they aren't held down by other foods. One way to anchor runny ingredients is by placing other foods on top of them. For example, you can angle sliced meat or vegetables against purees and mashed vegetables.
- Serve odd amounts of food. If you're serving small foods like shrimp, scallops, or bite-sized appetizers, always give guests odd quantities. Serving 7 brussels sprouts instead of 6 creates more visual appeal, and diners will also perceive that they're getting more food.
- Place food to create flavor bites. Essentially, flavor bites are forkfuls of food that combine all of the ingredients in your dish into one bite. Creating flavor bites is the perfect accompaniment to creative plating as it pleases both the eye and the taste buds.
- Don't overcrowd your plate. Be sure to never overcrowd your canvas, and keep it simple by focusing on one ingredient - usually the protein. Finding a focal point also ensures that the accompanying ingredients will play a complementary, supporting role.

3. Pay Attention to the Details

As you plate your dish, you'll also want to pay attention to the details:

- Think about color and contrast. One of the best-kept secrets to beautiful plating is paying close attention to the details. While your focus will obviously be on the protein, considering how the other elements of the plate create color and contrast is also very important.

  You can create a beautiful background for your plate by adding green vegetables or brightly colored fruits as accent points. Similarly, try to pair ingredients with complementary colors as this will further enhance your dish's visual appeal.

- Create height on your plate. Another way to catch your guests' eyes is to utilize the power of height. While compactly stacking ingredients isn't as popular as it was 5-10 years ago, creating a tall plate can go a long way towards enhancing visual appeal.
- You can also balance out taller ingredients by leaning long, flat items against them. For example, you can plate your steak on top of polenta and lean asparagus spears against them at a 45-degree angle.
- Use texture to enhance your dish. Finally, don't forget about texture. Contrasting a smooth vegetable puree with crunchy onion straws or topping a steak with crumbled blue cheese creates appealing texture combinations that are classic in high-end cuisine.

4. Design and Create with Sauces

Once you've plated your main ingredients, you're ready to top your dish with delicious sauces. Don't just pour the sauce carelessly all over the plate, though. Instead, think of your squeeze bottle or spoon as a paintbrush, and your sauce as a medium. Then, use them to enhance your plate.

One way to do this is to create accent dots on one side of your plate (while considering the rule of thirds) or by lightly drizzling sauce over the main ingredients so guests get a little bit of sauce in every bite.

5. Use Garnishes Purposefully

In the past, chefs casually threw a piece of kale and an orange slice onto every plate as it left their kitchen. However, these garnishes didn't add anything exciting to the dish, and few guests even ate them in the first place. Here are a few examples of smart garnishes and how to incorporate them:

- Choose edible garnishes. As you finish plating, remember that garnishes must be related to the dish and should always be edible. Ultimately, they're designed to enhance and complement the flavors of the entree you've created, not distract from them.
- Place garnishes purposefully. Similarly, never heap garnishes in one corner of the plate. Instead, disperse them thoughtfully in order to add color or texture. Also, avoid using unappetizing garnishes like raw herbs, large chunks of citrus, and anything with a strong odor. Lastly, make sure your garnishes are quick and easy to apply, so food still goes out piping hot.

Examples of Plating Styles

Here are examples of three popular plating styles: classic, free form, and landscape. To demonstrate them, we used filet mignon, potato puree, carrots, a demi-glace, a pea puree, a lima bean and pea blend, thyme, and fried leeks.

Classic Plating

1. Pipe the potato puree onto the plate using a pastry bag.

2. Place the carrots next to the puree using precision tongs.

3. Garnish the carrots with thyme using precision tongs.

4. Plate the steak using precision tongs.

5. Garnish the steak with fried leeks using precision tongs.

6. Drizzle the demi-glace around the plate using a spouted saucier.

7. Wipe the edges of the plate with a clean towel.

8. Finished classic plate.

Free Form Plating

1. Pipe dots of potato puree onto the plate using a pastry bag.

2. Slice the steak into three pieces using a chef's knife.

3. Plate the pieces of steak using precision tongs.

4. Place the lima bean and pea blend around the plate using a spoon.

5. Plate the carrots using precision tongs.

6. Place dots of pea puree around the plate using a large squeeze bottle.

7. Place dots of the demi-glace around the plate using a small squeeze bottle.

8. Garnish the plate with fried leeks using precision tongs.

9. Wipe the edges of the plate with a clean towel.

10. Finished free form plate.

Landscape Plating

1. Place dots of pea puree around the plate using a large squeeze bottle.

2. Paint the pea puree onto the plate using a brush.

3. Pipe the potato puree onto the plate using a pastry bag.

4. Plate the carrots using precision tongs.

5. Lean the steak against the puree and carrots using precision tongs.

6. Place the lima bean and pea blend around the plate using a spoon.

7. Drizzle the demi-glace around the plate using a spouted saucier.

8. Garnish the steak with fried leeks using precision tongs.

9. Wipe the edges of the plate with a clean towel.

10. Finished landscape plate.

How to Improve your Plate Presentation

Here are Steps to improve the visual presentation of your plates:

Step 1: Choose your Plates Carefully

Beautiful crockery is essential as it acts as the backdrop for your food. Play with colors, patterns and textures. In general use pale plates for bold vibrant colors so that the contrast is highlighted. Darker plates can be used for lighter colored foods. If you use patterned tableware, make sure that the overall look is muted enough so as not to distract your guests from their meal. Use large plates as you will have more room to present your food.

Step 2: Use Mini-dishes

When serving dishes such as stews or beef bourguignon, use mini-casserole dishes and ramekins for a fun look.

Step 3: Don't Serve Huge Portions

Of course, you don't want to starve the people at your table, but remember that the smaller your portions, the easier they will be to style. Also stick to 4-6 elements on each dish, any more and you risk overcrowding.

Step 4: Contrast Colors and Play with Symmetry

This will add energy and fun to your presentation. But remember not to repeat the same look too often, otherwise you'll be in danger of becoming monotonous.

Step 5: Position your Ingredients with Care

Imagine that your plate is the face of a clock: the centre piece of your dish (meat for example) should go at 6 o'clock. Place vegetables at 11 o'clock and perhaps starchy foods at 2 o'clock for slight asymmetry.

Step 6: Arm yourself with the Right Tools such as Pastry Cutters and Molds

Very helpful in giving shape to your dishes and to build layers.

Step 7: Use a Brush or Pipette

This will allow you to be more precise when creatively applying sauce: points, straight lines or zigzags the choice is yours.

Step 8: Piping Bags should be a Kitchen Essential

For precise assembly and plate dressing, a piping bag will be your best friend.

Step 9: Use Small Individual Sauce Boats

Really useful if you are serving a curry or a stew. This allows your guests to help themselves to as much or as little as they like, and it will be neater.

Step 10: Use Herbs, Spices, Edible Flowers or Citrus Zest for Decoration

By using edible flowers and micro herbs you can completely transform the look of your dish.

How to Design your Plate by Simple Geometric Shapes

Simple geometric shapes are the "skeleton" of plate design, the base upon which the flesh is added. Consider all the images which follow and how all plate presentations can be defined by simple geometric shapes: lines, arcs, circles, etc. Being able to "see" the underlying patterns (skeleton) of a layout make it easier to visualize the end result, and to play with variations of design before you ever put food to plate.

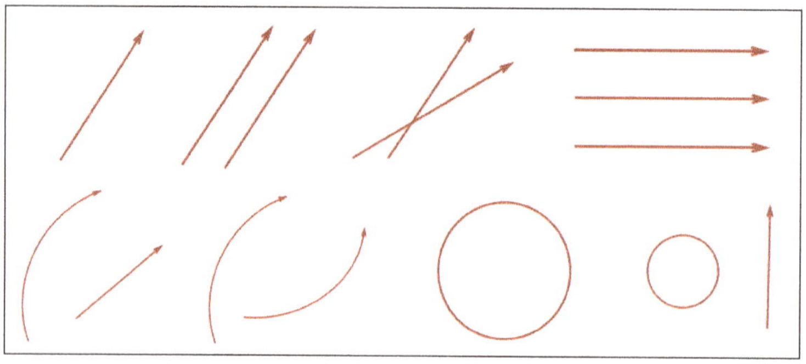

Single Line

These presentations takes full advantage of white space.

Chef Stephanie Alexander

Chef Alan Murchison

Two Lines

Two lines can be used in a variety of ways to create stunning presentations. Sometimes used as 2 parallel lines, somes crossing each other to form an "X".

Two Vertical Lines

Chef David Buchanan

Horizontal Lines

Horizontal lines on the plate are a great way to really feature the full details of what you are displaying because this perspective gives the fullest view of what's being presented.

Chef Daniel Humm

Canlis, Seattle

Two Lines – Crossing (Like an X)

What's interesting about this design matrix is that it is more interesting, more visually appealing, if the crossing point is off-center, especially if the starting and ending points of each line is also random (i.e. not side by side).

Chef Alain Ducasse

Chef Sven Elverfeld

Three Vertical Lines

This style works especially well if you are plating a trio.

Éric Gonzalez – Ivan Apostolov

Chef David Buchanan

Arcs

Arcs (affectionately called a "swoosh") add fluidity and a sense of motion to a presentation. If your design looks too "hard" or too "blocky" try adding an arc or two.

notice the 4 arcs

Chef Elie K

Arc and Line

A single arc and a single line. This design brings both fluidity (the arc) and stability or strength (the line) to a presentation. The center image is a horizontal line inside an arc.

Chef Kelly Liken

Chef George Tannock

Chef David Buchanan

Circle and Line

Similar to Arc & Line, this offers both a soft and a hard component to the plate. Typically the circle is off-center to the left.

Chef Chris Nugent

Chef Chris Nugent

Chef Peter Michael L

Structured Chaos

Some plates look like chaos, but a closer look reveals structure. This is one of the hardest plating styles to successfully pull off if its wrong then all you have is a mess on a plate. Balance, white space, and a focal point is the key.

Chef Jan Verhelst the vertical line on the left anchors this design.

Chef Yann Bernard Lejard the master of artistic "splat"

Triangle

The classic 3 component (starch, veg, protein) design: 10 o'clock, 2 o'clock, 6 o'clock. But also used for more avante garde presentations.

Chef David Buchanan

Centered Circle

A classic presentation is to center everything, making a tower in the center of the plate, stacking all the components, and then putting a "mote" of sauce around it all. This presentation style also gives great height (elevation) to the design.

Chef Dean Shinagawa

Chef Dean Shinagawa via Seattle PI

Off-center

A more modern approach is to place items slightly off-center. Being a little off-center creates tension and interest for the eye.

Chef Yann Bernard Lejard

Chef Thomas Keller

Chef Ahmed Fatheen

Complexity

The plate design below is beautifully intricate yet somehow appearing to be simple. Complex plate designs are difficult to pull off for multiple reasons. First, too much on the plate can quickly turn into confusion. It takes a skilled eye to make complexity aesthetically balanced. Second, intricacy takes time to plate correctly do you have time to plate this dish correctly during service? Third, very few people have the eye and the hand to plate complex designs correctly so you will probably have to create every plate yourself during service.

Chef Gianfranco Chiarini

Breaking the Rules – Decorating the Rim

The old rule of plating was to keep everything off the rim. The rim acted as a frame for the plate and the Chef would yell at you if you had even a speckle on the rim. Today, the rim is often still considered to be a border but it is a border to be played with.

Restaurante 100 Maneiras

Chef Vicky Lau

Plating Tools

Having the proper food presentation and plating tools is essential to high-quality plating. Here are a few items you should be sure to purchase if you don't already own them.

Decorating Brushes

As one of the most important products in any chef's toolkit, decorating brushes have a variety of applications. You can use them for both detailed line work and broad strokes as you apply sauces, or when plating purees and coulis beneath meat or vegetables.

Garnishing Kits

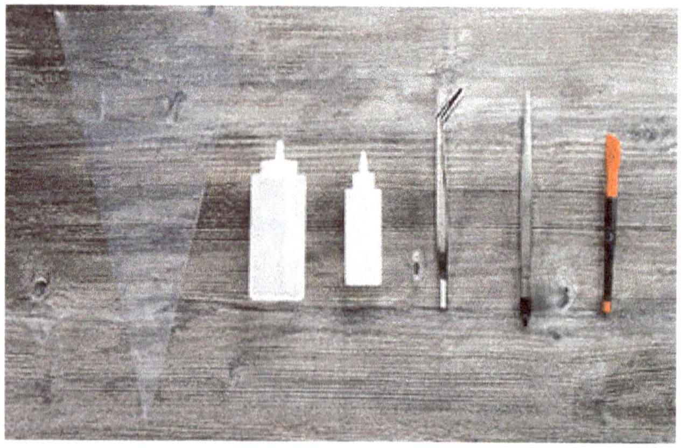

Garnishing kits come with everything you need to garnish all of your signature dishes, including plating wedges, tongs, squeeze bottles, and brushes.

Molds

Molds are also very important when plating food. By cutting ingredients to a specific shape and size, you'll provide visual appeal and keep your plate tidy. Ring molds also help you develop height and structure when stacking ingredients.

Plating and Precision Tongs

Last but not least, you'll want to have precision tongs on hand for placing garnishes or small, delicate foods. Many tongs also feature micro-serrations for improved grip and stability.

Plating Wedges

Plating wedges come pre-cut with flat, round, or pointed edges and are perfect for smearing sauces and other soft ingredients into designs on your plate.

Shavers

Shavers work well when shaving or grating chocolate, hard cheeses, or soft vegetables on top of your finished creations.

Spoons

You'll also want to have a variety of spoons on hand. Saucier spoons help you drag smears of sauce across your plate, and you can also choose a utensil with a tapered bowl that's perfect for drizzling and pouring. Additionally, slotted spoons quickly separate solids from liquids as you complete your presentation.

Squeeze Bottles

Squeeze bottles are ideal when applying sauces or aiolis to your finished plate. Many of these items come with adjustable, precision control tips that allow you to apply the perfect amount of product.

How to make a Tomato Rose

A tomato rose is an elegant way to dress up your dishes at your next family gathering or special event. Even though it may look difficult to create, almost anyone can make a tomato rose garnish. Follow these steps and you'll create your very own tomato rose to impress your guests at your next event.

Step 1

To make a tomato rose garnish you will need:

- Sharp paring knife
- Cutting board or other flat surface
- Medium size tomato (firm to the touch)
- Greens for added detail - mint, lettuce, sage, or other type of leaf

The key to successfully learning how to make a tomato rose is by using the right tool. A sharp paring knife is essential, as it allows you to control the depth and direction of your cuts. A dull blade will make it virtually impossible to create a pretty tomato rose.

Step 2

The right tomato is important. Look for these qualities:

- Firm and smooth.
- No marks or blemishes. (Blemishes and marks make it more likely that the peel will break during the cutting process).
- Good red color (although other shades of tomatoes can make beautiful roses as well).

Buy vine ripened tomatoes because they seem to be a prettier shade of red and of a good size. Any size tomato can be used to make a rose.

Start at the bottom. Turn the tomato upside down so that the stem side is in the palm of your hand and the bottom is facing you. Starting about 3/4" from the center, cut a strip of the skin across the bottom that is about 1" wide and about 1 ½" long. Keep the knife blade just under the skin of the tomato and be sure not to cut through the peel completely, as the end goal is one continuous peel from the entire tomato. At the end of the cutting process, this wide piece is going to be the base for your rose, which will keep it from unraveling.

Step 4

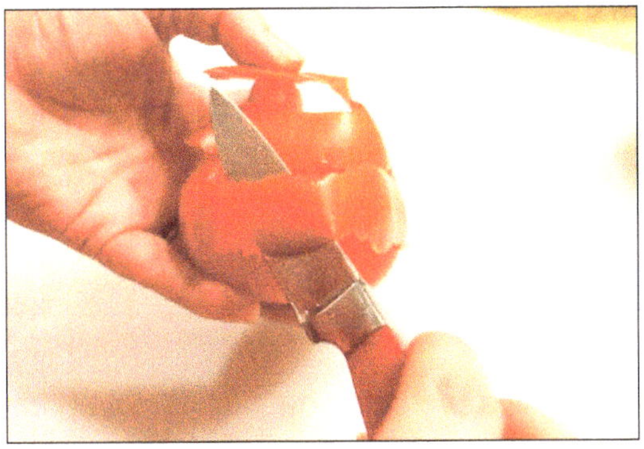

This is probably the trickiest part. While cutting, slowly turn the tomato and without detaching the base, begin peeling the tomato in a single strip, about 1/2" wide, in the same manner you would peel an apple.

Step 5

Cut around the tomato keeping the cut the same depth, close to the skin of the tomato. As you get near the original cut aim to cut under and below it. Continue cutting around the tomato, keeping the skin all in one piece. Gradually make the strip of tomato skin slightly narrower as you get closer to the core.

Step 6

Finally cut the skin from the tomato. The skin should be in a single piece with a wide end and a narrow end.

Step 7

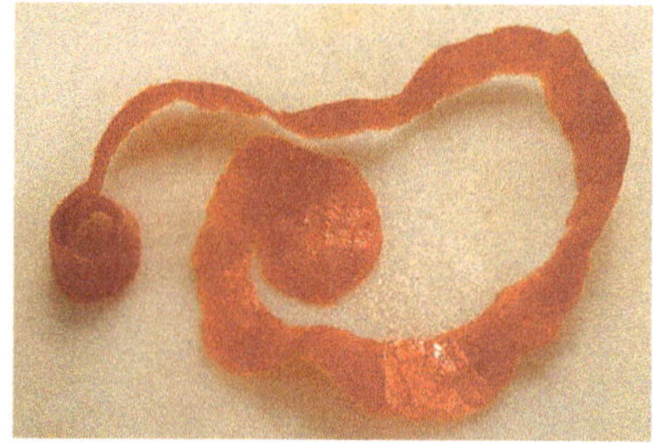

Carefully lay the peel out, fleshy side up, on your work surface the best you can. You can start at either end for rolling your roses.

Starting at the base: Make the outside ring by loosely positioning a layer of the peel around the outer edge of the base.

Starting from the core: Using your fingertips, begin rolling the rose, keeping it tight and curling it around itself.

Step 8

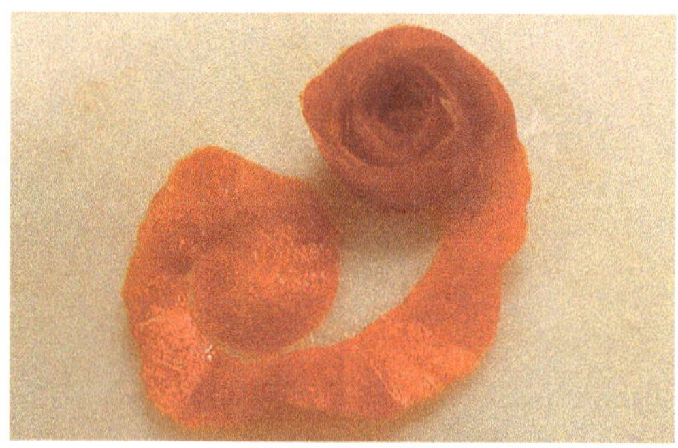

From base: Continue twisting the layers into the center of the rose loosely enough that you can still add to the center.

From core: Continue rolling the peel around itself. This is easiest done with the peel set down on the cutting board and curling each layer around the outside.

Step 9

Finished. One, very easy, tomato rose.

There is a very slight difference to the center of the rose depending on which way it was wrapped. The rose started from the base has more of a twisted in center due to trying to fit the last part of the peel to the middle, as compared to the rose started from the middle which has a more uniform appearance.

Step 10

Finishing touches: To add a little extra color to your tomato rose, add some greens to form the rose leaves when you place it on a serving tray. You can use parsley and spinach leave for your roses, but other great greens include: mint, sage, lettuce, and kale. Don't hesitate to be creative and try different ideas, even purple leaves make a beautiful presentation.

Step 11

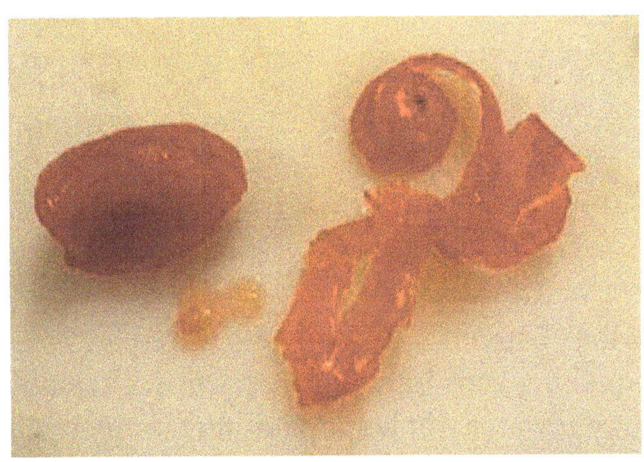

As mentioned before, any size tomato will work, if you have the skill and the desire to try it. A serrated paring knife works better for the smaller size because it cuts slower and allows for more precision narrow cutting.

Step 12

Finished Tomato roses to give that added beauty to any display.

Step 13

Storing: Tomato roses can be covered and refrigerated for several hours or overnight. But, if you are making your tomato rose garnish ahead of time, keep it fresh by soaking it in very cold water. The cold water will help maintain freshness and enhance the appearance when used later.

How to make Watermelon Checkers

This is a great way to serve cheese and fruit, and even the kids will enjoy eating their healthy food. This watermelon and cheese checkerboard is a wonderful idea for a child's party or even for placing out on the buffet table. Here is how to prepare the checkerboard in just a few minutes.

Steps

1. Buy hard cheese and watermelon. The cheese can be yellow or orange, whatever is available in your area.

2. Prepare the watermelon pieces. Cut slices of watermelon 1/2 inch or 1.3 cm thick. Use a ruler if you're not able to gauge this easily.

3. Turn the slices on their side and cut into strips of equal width.

4. Cut into equal squares. Do this as evenly as you can. The exact square size is up to you and will be determined by the amount of watermelon and cheese you have, or the plate size.

5. Set to one side for arranging shortly.

6. Prepare the cheese. Slice the cheese 1/2 inch or 1.3 cm thick.

7. Cut the cheese into even squares the same size as the watermelon squares.

8. Arrange the pieces into a checkerboard formation on a suitable serving plate.

- This is a good time to remove pieces that are lopsided or of the wrong size – cut more if needed, or simply reshape existing pieces.
- Also remove any obvious seeds from the watermelon pieces, if you haven't used seedless watermelon.

9. Prepare the checkers. Use a very small cookie cutter or a small shooter glass turned upside down to cut out the checkers from the watermelon. Cut slices thinner than for the checkerboard squares.

- Push all the way through to ensure that the checker piece comes away clean from the watermelon flesh.

10. Repeat the cookie cutter process with the cheese to make cheese checkers.

    - As with the watermelon checkers, cut the slices thinner than cheese squares for pushing the cutter through.

11. Arrange the checkers on the checkerboard and firm up the entire checkerboard. Place the watermelon checkers onto cheese pieces, and cheese checkers onto watermelon pieces, as this makes them easier to see and it looks more effective. Your watermelon and cheese checkerboard is ready for serving.

    - It's a good idea to provide toothpicks to enable guests to pick up the cheese and watermelon pieces easily.

How to make an Apple Swan

Decorative items on a dining room table can accent a meal and contribute to a chic and sophisticated atmosphere. An apple swan is an edible decoration that's relatively easy to create with just an apple and a knife. By following the correct procedures and adding your own creative flair, you can create an apple swan that will impress dinner guests.

Method 1. Creating the Body

1. Purchase a large apple. Purchase an apple that stands up straight from the grocery store or farmer's market. You can get any variety of apple, but make sure that it is as large as possible because it will make it easier to create your swan decoration.

2. Cut through the center of the apple at a slight angle. Stand the apple upright and use a sharp knife to cut through the core of the apple at a slight angle. Cut it by placing your knife 1/4 of an inch (0.635 cm) away from the middle of the apple, and make a long vertical cut down the center. Make sure to go through the core itself. Remove the exposed seeds with the pointed end of your knife and set them aside for later.

- The two seeds will be used as eyes later on.

3. Place two butter knives on both sides of the apple. Take the larger side that has the base and

top of the apple and place it flat on the cutting board. Place two butter knives perpendicularly at the base and the top of the apple. These knives will act as bumpers and prevent you from cutting through the entire apple.

4. Make a vertical cut 1/2 of an inch (1.27 cm) to the right of the apple core. Use a knife and make a vertical cut until your cutting knife hits the butter knives that you placed as a bumper.

5. Make a horizontal slice to cut out a wedge from the apple. Use the knives as a guide again and make a horizontal cut until you reach the other cut that you just made. The piece of apple that falls out should look like a wedge.

6. Cut out a wedge on the other side of the apple. Repeat the last two steps on the other side of the apple. This should make it so that your apple now has two wedges cut out of both sides with a thin body in the center of the apple.

Method 2. Making the Wings

1. Lay two knives on both sides of your apple wedge. Set your apple wedge on its flat end between two knives. The knives should be at the top and bottom of your apple wedge.

2. Make a quarter of an inch (0.635 cm) vertical cut through the wedge. Make another vertical cut a quarter of an inch (0.635 cm) from the flat side of the wedge of apple. Continue to cut until the butter knives stop you.

3. Flip your wedge over and make another vertical cut. Instead of making a horizontal cut and risk cutting yourself, flip your apple on its other flat side and create another vertical cut. This should cut out another, smaller wedge of apple from the larger wedge that you cut.

4. Repeat the process two more times on the smaller apple wedges. Set the smaller piece of apple between the butter knives and create another wedge out of it. Out of your smaller wedge, repeat the

process until you have three or more different sized wedges. These will act as the different layers of your swan's wings.

5. Stagger your pieces of apple to create wings on the body. Push out the wedges of apple so that they are staggered and appear like wings. Each wedge should show a little bit of the wedge under it to create a wing-like effect.

6. Cut another wing out of your other apple wedge. Repeat the steps using the other large apple wedge that you had cut out of the other side of the apple. Continue repeating the steps until you have two staggered wings.

Method 3. Making the Head

1. Cut a half inch slot (1.27 cm) out of the middle of the body. Use a small knife to cut out a small half inch (1.27 cm) wide rectangle. The cut can be about 3/4-inch (1.905 cm)deep.

2. Cut a slice of apple out of the remaining side of your apple. Set two knives next to the other half of your apple and make a full horizontal cut through the bottom of the apple. This should create a thin slice of the apple that you can use for your swans head and resembles a heart.

3. Take the slice of apple and make two diagonal cuts to form a V. Take the flat slice that you just cut and make two diagonal cuts near the top right of the slice of apple. This will start to form the head. Try to visualize what a swan's head looks like, or use reference photos from the web so that you can create the best swan head and neck.

4. Make one straight cut on the bottom for the base. Finish your swan's head by making one final cut on the side of the apple slice, below the V that you just cut out. You can make other cuts and refine the look of the swan's head as well. Round off harder angles with a small knife around areas like its neck and head to give it a more realistic look.

5. Press the apple seeds where the eyes should go. Take the apple seeds that you set aside and put them into where the eyes should go on your swan's head. Use a butter knife to press them into the pulp of the apple, so they stick.

6. Drop the neck into the slot that you created. Take the swan head that you created and place it into the rectangular slot that you made earlier. If you're having trouble keeping the swan's head in the notch that you cut out, you may have to cut a little bit deeper to accommodate the size of the neck.

How to make a Fruit Bouquet

Fruit arrangements are festive, eye-catching centerpieces. These edible bouquets are appropriate for every occasion, from baby showers to Fourth of July barbecues, and they make excellent gifts. Instead of serving a fruit salad at your next potluck or giving your partner a box of chocolates next Valentine's Day, treat the people in your life to a delicious, attractive, and homemade fruit bouquet.

Part 1. Selecting your Ingredients and Materials

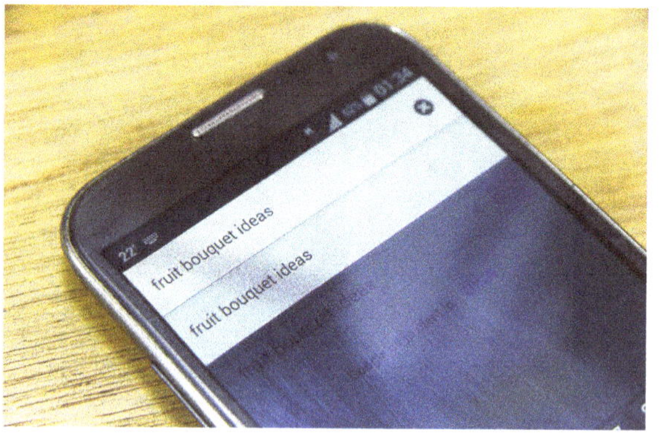

1. Search for inspiration. Prior to purchasing your fruit, it can be helpful to have a plan or a general idea of what the finished product will look like—especially since the possibilities are endless. When you have a general idea of the size and design, you be able to more accurately estimate the type of fruit you need to purchase and how much of each kind you need to buy.

- Make a bouquet with nothing but strawberries. Craft a more complex bouquet out of all different types of fruits, like pineapple, watermelon, and muskmelon. Cut your fruit into unique shapes, such as stars, balls, and animals. Transform your fruit into a replica of a baby, bride, or one of your favorite characters.

- If you are not a visual person, it may be helpful to conduct a quick internet search for examples of fruit and floral arrangements. Allow these images to inspire your vision for the fruit bouquet. You may find or draw a design that roughly resembles a floral display, and then have fun imagining which fruit will replace the flowers.
- If you're not a planner, feel free to design the bouquet as you are creating it.

2. Select your fruits. When you choose fruits, there are several factors to consider. Does the fruit brown quickly? Is the fruit hardy enough to stay on a skewer? Do the colors of the fruit complement each other and or your theme? Popular fruits for edible arrangements include:

- Berries: strawberries, raspberries, blueberries, and blackberries
- Melon: watermelon, cantaloupe, and honeydew
- Citrus fruits: oranges, mangos, and pineapple
- Grapes

3. Avoid fruits that brown. Fruit bouquets are edible centerpieces. In order to keep your arrangement looking fresh and appetizing, do not include fruits that brown quickly. These fruits include:

- Bananas
- Pears
- Apples

4. Consider adding sweets to your arrangements. Fruit pairs really well with sweet treats. You can elevate your fruit arrangement to a delectable dessert by dipping a portion of the fruit in milk, white, or dark chocolate. You may also wish to add mini-marshmallows or puffed cereal to your skewers.

5. Choose a base for your arrangement. Fruit bouquets are displayed in a variety of containers, including vases, cups, pots, and baskets. You can even use other fruit as a base, such as a melon or apple. Select a container that is appropriate for your party or gathering.

Part 2. Prepping the Fruit and Making the Kabobs

1. Wash and dry the fruit. Prior to creating the individual stems of the arrangement, you should

thoroughly wash and dry all of your fruit. Rinsing your fruit with water will remove dirt, grime, pesticides, and bacteria from the surface of the produce.

- Individually rinse the melons, oranges, and mangos.
- Place one type of berry or grape into a colander at a time. Rinse the fruit and the spread it out on paper towels or kitchen towels to dry.

2. Prep and cut the fruit. It is important that your edible bouquet is consumer friendly. Remove all inedible materials from the fruit, such as stems, seeds, and leaves.

- Hull the strawberries. Use a knife to remove the stem and hull from the strawberries.
- Remove the seeds from your muskmelon and honeydew.
- Remove the stems from your grapes and berries.
- If you are using fresh pineapple, cut off the ends of your fresh pineapple and then use a knife to remove the sides of the pineapple from the fruit.

3. Cut the fruit. Edible arrangements feature fruit that has been cut into both festive shapes, such as flowers, animals, and even letters, as well as traditional shapes, such as wedges or balls.

- Instead of creating the festive shapes by hand, you can use cookie cutters to produce uniform, professional-quality cutouts. Select a variety of sharp edged cutting cutters that match the theme of your event.

- Create pineapple cutouts. Cut the fruit into ¾ inch thick slices. Place the sliced pineapple in the freezer for one hour. Remove the fruit from the freezer and use the cookie cutters to transform the pineapple into fun shapes.
- Create melon cutouts. Slice the melon into ¾ inch thick slices. Cut fun shapes out of the melon with cookie cutters.
- Slice the melon. Sliced melon is a great gap filler. Leave the rind on the melon. Cut it into slices or wedges. Instead of a traditional knife, you can use a scalloped garnish tool to cut the melon.
- Ball the melon. Use a melon baller to create uniform balls of fruit.
- Cut oranges, mangos, apples, pears, or kiwis into slices or wedges.

4. Dip the fruit in chocolate. To appease those with a sweet tooth, you may wish to dip some of the fruit in chocolate.

- Melt the chocolate in the microwave or on the stovetop in a double boiler.
- Use toothpicks to dip your fruit into the melted chocolate.
- Set the dipped fruit on a sheet of wax paper.
- Once all the items are dipped, move the fruit to the refrigerator and allow the chocolate to set for 15 minutes.

5. Assemble the skewers. After all of your fruit is washed, prepped, and cut, the fun of creating designs out of the fruit can begin. Don't stress about this part of the process—there is no right or

wrong way to assemble a fruit skewer. You may use any size and type of food safe skewer, from a toothpick to a wooden kabob stick; you have the freedom to arrange the fruit in any order or combination you choose. As you create, just make sure to leave enough space on the end of the skewer to stick into the base. Common combinations include:

- Alternating colors of grapes and or berries.
- Pineapple and melon flowers. Slide a melon ball, marshmallow, or strawberry to the top of the skewer (the non-pointy end). Next, slide a few different sized melon and or pineapple flower cut outs onto the skewer. Fill the remaining space with grapes or berries.
- Melon wedges.
- A single chocolate covered strawberry or a chocolate covered strawberry followed by a row of mini-marshmallows.

Part 3. Putting Together the Fruit Arrangement

1. Prepare the base. Insert a piece of floral foam, Styrofoam ball, or a head of lettuce into the container—all of the skewers will be stuck into this item. Use toothpicks to attach decorative tissue paper or dark leafy greens to the surface of the floral foam, Styrofoam ball, or head of lettuce.

- If you use floral foam, you can place parsley on top of the floral foam to fill in any gaps that can be seen between the skewers.
- Do not let the florist's foam come into direct contact with the food. Wrap the foam completely in plastic wrap.

2. Insert the skewers into the base. When you are arranging fruit, or flowers, your goal is to create

an even and aesthetically pleasing bouquet. Insert your tallest skewers of fruit into and near the center of your base. As you work your way to the outer edges, the skewers should decrease in height.

3. Assess the arrangement. Periodically step back and evaluate the arrangement. Remove any skewers that seem out of place. Fill in all of the gaps with additional skewers. Continue the process until you are satisfied with the arrangement.

- You may need to skewer additional fruit to fill in the gaps.

4. Store and or transport the arrangement. After you have assembled the arrangement, wrap the fruit bouquet in plastic wrap or a cellophane gift bag. Transfer the arrangement from your workspace to the fridge. Pull the fruit bouquet out of the fridge 5 minutes prior to the party or moments before you intend to deliver it. If you are transporting the arrangement in a vehicle, place the edible centerpiece in a cardboard box and fill the empty space with bags, clean towels, or tissue paper.

How to use Nasturtiums in Food

The nasturtium (Tropaeolum majus) is a hardy annual that grows by climbing and sprawling. It produces beautiful flowers that can be put to great culinary use, and the leaves are also edible.

Method 1. Preparation

1. Use any part of the nasturtium. The leaves, seed pods and flowers are all edible. The flavor is a cross between mustard and slight sweetness.

- Flowers should be harvested just as they open. Smaller flowers are usually best for eating, while larger flowers are better as garnishes or for being ripped apart and chopped or minced.

Method 2. Adding to salads

2. Use the leaves in a salad. The leaves have a hot, peppery taste. Simply chop them up and add to the salad.

- The leaves make an ideal substitute for watercress.

3. Throw the flowers into the salad. The flowers come in a range of colors from gold and orange to red. These can make a beautiful feature in a salad, with the bonus that they're edible too.

Method 3. Using Nasturtiums as Caper Substitutions

1. Use the seed pods (green) as a good substitute for capers. Use them in any dish where you'd usually use capers, such as on pizza or in salads.

- They're best pickled first.

Method 4. Making Nasturtium Condiments

1. Use a combination of leaves and seeds. Blend them into butter or soft cheese for an added spicy flavor.

2. Add nasturtium petals and leaves to oils, dressings or vinegars. Mince them prior to adding.

Method 5. Adding as a Garnish

1. Add the flowers to savory dishes and desserts, to given them a touch of elegance. For example:

- Brighten up breakfast.

- Make appetizers even more appetizing.

- Decorate a first course.

- Make main course tasty.

- Enjoy them with a savory-inclined dessert.

Method 6. Stuffed Nasturtiums

These little delicacies are perfect for afternoon tea finger nibbles.

1. Select large, whole flowers in good condition. While not essential, it's more effective to choose a variety of colors for variation.

2. Choose a suitable stuffing. Any soft cheese, dip or guacamole will make an ideal stuffing.

3. Put teaspoonfuls of stuffing inside the flower until about three quarters full. Gently fold the petals back around the stuffing, the stickiness of which should hold the petals in place.

4. Serve with Melba toast or crackers. The stuffed flower goes on top of the toast or cracker for ease of eating and completes the texture perfectly.

    - Alternatives to toast or crackers include slices of cucumber, sticks of celery or slivers of capsicum/bell peppers.
    - The stuffed flower can be eaten alone if desired.

Method 7. Making a Nasturtium Omelette

1. Cut the runner beans into thin slices. Add to a saucepan of boiling water and boil for 4 minutes. Drain and set aside.

2. Crack the eggs into a mixing bowl. Pour in the milk and beat together.

3. Using the back of a fork, crush the nasturtium seeds. Add them to the egg mixture. Also toss in the leaves and petals.

4. Season to taste with salt and pepper.

5. Add the butter to the frying pan. Melt over gentle heat.

6. Pour in the egg and nasturtium mixture. Add the boiled beans and stir quickly. Then allow to cook gently until the omelette sets.

7. Serve. Sprinkle grated Parmesan cheese over the omelette and serve. Garnish with nasturtium petals.

How to make Fondant Flowers

Have your cake and decorate it too with beautiful blooms made of fondant. Fondant is a type of edible frosting that's thicker than your usual icing, so it's perfect for sculpting gorgeous cake decorations. To try it yourself, shape roses or carnations by hand or, if you want an easier method, use a mold to make the prettiest petals.

Method 1. Shaping Fondant Carnations

1. Roll the fondant into a flat sheet that's 1⁄8 in (0.32 cm) thick. Use a rolling pin to flatten the fondant, applying gentle pressure as you move the pin back and forth to make sure the entire fondant is an even thickness.

- You can use any color fondant you want for your carnation.
- To prevent the fondant from sticking to the rolling pin or work surface, dust both the surface and the rolling pin with a little cornstarch before you begin, or apply a thin layer of shortening.
- To change the color of your fondant, squeeze a few drops of gel coloring onto the fondant. Then massage it in, kneading the fondant to spread the color evenly throughout.

2. Cut out 5 circles in different sizes. Use fondant cutters or cookie cutters to form your circles from the flattened fondant. Start with the largest circle, which will be the base of your flower. Then make a slightly smaller circle, continuing to decrease the size for each of the remaining 3 pieces.

- Adjust the diameter of the circles as needed to create smaller or larger flowers.
- For fuller flowers, cut out more than 5 circles.
- If you don't have fondant or cookie cutters, you can use anything that has a circular shape, like the bottom of a drinking glass. Press it into the fondant and cut around it with a knife.

3. Roll a toothpick around the edges of each circle to create ruffles. Press the length of the toothpick lightly into the fondant. Rock it back and forth, moving it around the perimeter of the circle. This causes the fondant to curl up into a ruffled edge. Repeat on all 5 circles.

- Be careful not to press too hard so you don't cut through the fondant with the toothpick. You just want to make an indentation.
- Some bakers choose to lay their fondant on a foam shaping mat during this step. It provides a little cushioning for the flowers to prevent them from ripping or sticking to the counter.

4. Press the largest circle into a fondant forming cup. This will be the bottom of your flower. Set it in the fondant forming cup so that the edges curl up slightly, giving you a nice curved shape instead of a flat flower.

- You can use a small ramekin or even an empty egg carton instead of fondant forming cups.
- If you're worried about the fondant sticking to the cups, sprinkle some cornstarch on the inside of the cups first.

5. Brush water onto the circle with a small paintbrush. The water will act like glue to hold the flower together. Use just enough water to dampen the top of the fondant. Avoid soaking it through or using so much water that it puddles in the center of the circle.

- Don't use a paintbrush that has been used for paint or any other toxic substances. Buy a brand new one that hasn't been contaminated.

6. Place the 4 other circles on top, brushing water in between each layer. Arrange the 4 circles in decreasing size order so that the smallest circle is the last layer. Use a dab of water between each of the fondant circles to "glue" them together.

- Press down lightly on each layer as you set it in the fondant cup to adhere it securely to the previous circle.

7. Let the flowers dry for 24 hours before using them. Place the flowers in a dry, room temperature spot to harden. The drier the flowers are, the better they'll hold their shape.

- Store any flowers you aren't using right away in an airtight container in the fridge for up to 2 weeks.
- Ideas for Decorating Your Fondant Flowers.
- Stick a shimmery sugar pearl in the center of the flower.
- Paint designs on the petals with a small paintbrush dipped in food coloring.
- Add green fondant leaves to the base of your flower.
- Use petals in all different colors for a whimsical effect.

Method 2. Making Fondant Roses

1. Roll the fondant into a rectangular shape that's 1⁄8 in (0.32 cm) thick. You want the fondant to be thin enough that the flowers don't look clunky, yet thick enough that it won't rip when you're forming the petals. Use a rolling pin with gentle pressure to evenly spread out the fondant on a flat surface.

- Spread a layer of shortening or sprinkle cornstarch onto your surface if you want to prevent the fondant from sticking.
- Use pink or red fondant for realistic roses. However, you can use any color you'd like.
- Trim the edges with a knife to create your rectangle, if necessary.
- Make Your Own Marshmallow Fondant

Ingredients:

- 1 10 oz (0.28 kg) bag of mini marshmallows
- 2 to 3 tablespoons (30 to 44 ml) of water
- 4 cups (950 ml) of powdered sugar

Directions:

- Melt the marshmallows and the water in the microwave, stirring every 30 seconds.
- Blend the powdered sugar into the marshmallows with an electric mixer or a spoon.

- Knead the dough on a flat surface for about 7 minutes until it's shiny and soft. If it's too dry, add 1⁄2 tablespoon (7.4 ml) of water at a time.
- Wrap the ball of fondant in plastic wrap and place it in the fridge for up to 2 weeks.

2. Cut a strip of fondant that's 1.5 to 2 in (3.8 to 5.1 cm) wide. The length of the strip doesn't matter but the width shouldn't be more than 2 inches (5.1 cm) or the rose will be too chunky. A butter knife will be sharp enough to slice through the fondant.

- Cut as many strips as you want roses. Each strip makes 1 rose.

3. Fold the strip in half lengthwise. Line up the long edges of the fondant to make sure you fold the piece exactly down the center. Don't smash the 2 sides together. Just rest them on top of each other so you can still see the crease between them.

- If you're having trouble folding the fondant, you may have a piece that's too thin. Re-cut a wider strip.

4. Roll the strip up, pinching the bottom to form a cone shape. Keep the open edge facing up.

Squeeze the bottom of the folded strip between your thumb and forefinger as you roll so it tapers down into a slight point.

- The tighter you roll your strip, the smaller your rose will be.
- If you wind up with extra fondant at the bottom, simply pinch it off with your fingers.
- Brush a little bit of water along the strip before you begin rolling if you want the rose to hold its shape better. The water acts as glue.

5. Set the roses in a room temperature spot to dry for 24 hours. Don't use the roses immediately to decorate a cake or they'll fall apart or become misshapen. Letting them dry first will harden them into their flower shape.

- If you aren't using your roses right away, store them in the refrigerator in an airtight container for no more than 2 weeks.

Method 3. Creating Fondant Flowers with a Mold

1. Dab cornstarch into each of the molds with a small paintbrush. The cornstarch will prevent the fondant from sticking to the mold. Use a brand new paintbrush that hasn't been used before, so there aren't any harmful paint chemicals on it.

- Be sure to coat every little crevice in the mold.
- If there's a lot of excess cornstarch in the mold, flip it over and tap it gently to remove the extra powder.

2. Break off a piece of fondant that's about the same size as your mold. If you're using a very small flower mold, you'll need a smaller piece. Don't worry if it's not exact; you just need enough fondant to at least fill the mold.

- If you want, you can roll the fondant out into the size of your mold. It may make it easier to press the fondant evenly into the mold later on, but it isn't necessary.

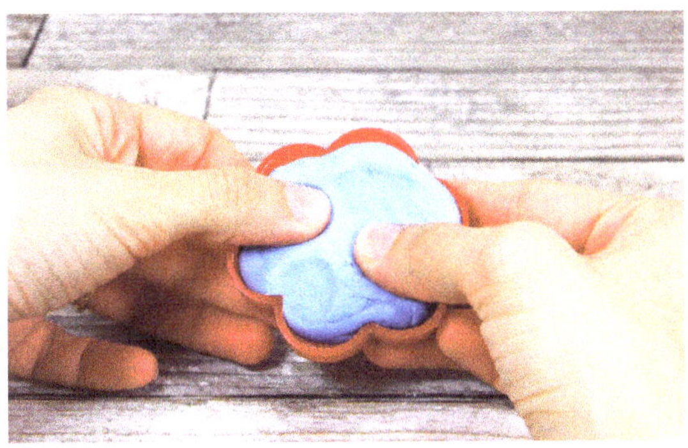

3. Press the piece of fondant into the mold, filling it completely. Use your fingers to firmly push down on the fondant, making sure you get it into every nook and cranny. It's okay if your fondant spills out over the mold. It's better to overfill it than underfill it.

- If your piece of fondant doesn't fill the entire mold, break off more fondant and press it on top of the original piece.

4. Shave off any extra fondant from the mold using a knife. Use a sawing motion to work the knife carefully along the top of the mold, pulling away the excess fondant with your fingers as you go. This should smooth out the fondant inside the mold.

- If you don't flatten the top of the mold (which will be the bottom of the flower), you'll have trouble sticking the finished flower to a cake or cupcake.
- You can use an icing spatula instead of a knife.

5. Place the filled molds in the freezer for 20 minutes. This hardens the fondant so that it's easier to remove from the molds. Set the flowers on a shelf where they'll sit flat and won't be disturbed.

- Set a kitchen timer or use the clock app on your phone to keep track of the time.

6. Remove the molds from the freezer and pop the flowers out of the mold. They should come out easily. Work quickly so that the flowers don't warm up in the molds, making it more difficult to get them out cleanly.

- Be careful with the flowers when you're removing them. They aren't completely dry yet, so too much force could alter their shape.
- If the fondant won't come out of the mold, leave it in for a little longer before trying again.

7. Let the flowers dry for at least 24 hours. Set the flowers in a room temperature spot to harden. If it's too warm, the fondant won't dry as well, or the flowers will lose their shape.

- Keep any flowers you aren't using immediately in an airtight container in the fridge for up to 2 weeks.

How to make Butterflies for Cake Decorating

Adding edible decorations to the top of a cake can create or accentuate the theme or mood to any event. Whether it is for cupcakes, cake pops, or a full sheet cake 3-dimensional chocolate or candy butterflies can create a beautiful, finished look without training, special equipment, or time.

Steps

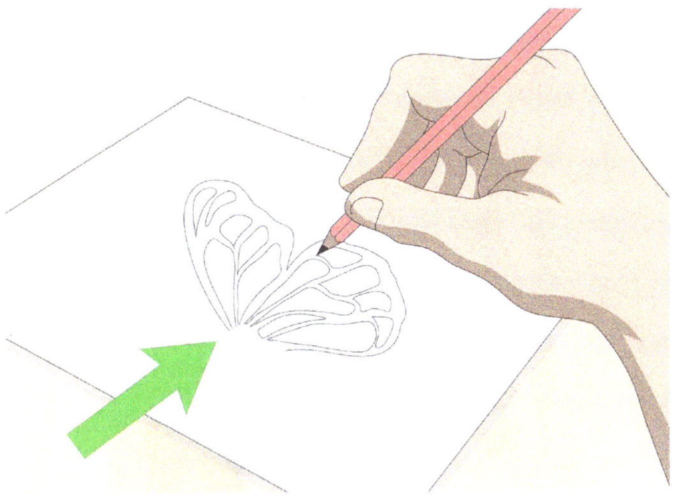

1. Create a wing template for your butterflies on a sheet of paper.

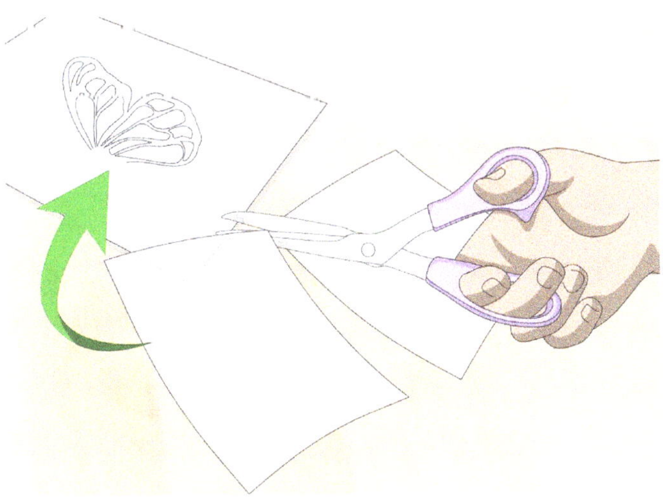

2. Cut squares out of the wax paper or parchment paper for each of the butterflies.

   - Make sure that you will have plenty of room to handle the wax paper or parchment for after you pipe your butterfly onto the wax paper or parchment.

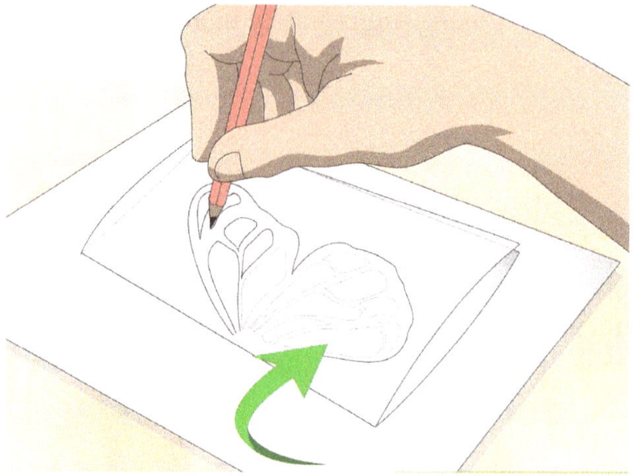

3. Fold the wax paper or parchment in half and trace your template pattern to the back side, using the fold as the body of the butterfly.

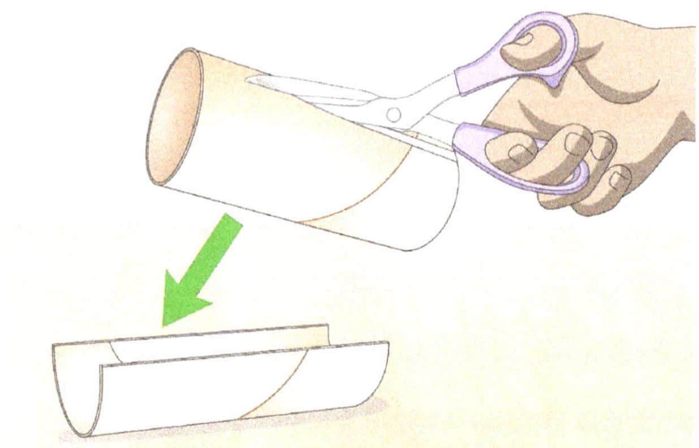

4. Cut an empty towel roll down one side

5. Fold a piece of thin cardboard or card stock in half so it opens up to 90 or more degrees and place it inside the split towel roll so the folded cardboard or card stock sits within the split towel roll.

6. Melt the chocolate or melting candy wafers in a bowl.

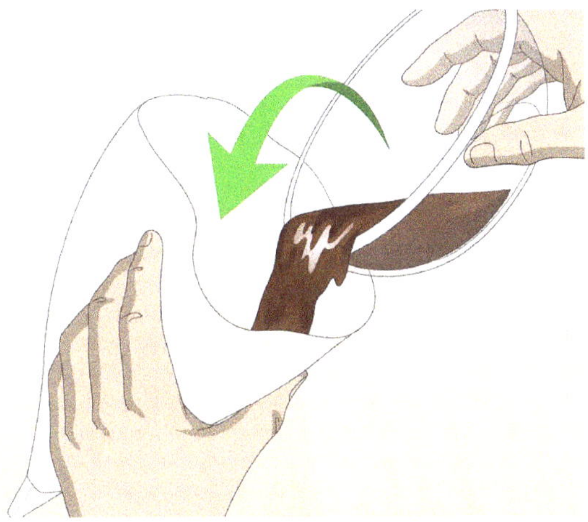

7. Fill your piping bag with the chocolate or melted candy wafers.
    - If you do not have a piping bag, cut a tiny hole in the corner of a plastic bag.

8. Pipe the melted chocolate or candy on the wax paper or parchment using your template as a guide.

9. Pipe a thicker line of chocolate or candy down the inside fold to connect the two wings and form the body and antennae of the butterfly.

10. Allow the butterflies sit for a few minutes until the melted material starts to set up but is not yet hard.

11. Lay the butterfly on the cardboard or card stock so that the body of the butterfly follows the line of the fold.

12. Place everything in the refrigerator or freezer and allow this to cool for an hour or until hard.

13. Carefully remove the chocolate or candy from the wax paper or parchment paper and place your butterfly on your cake.

Permissions

All chapters in this book are published with permission under the Creative Commons Attribution Share Alike License or equivalent. Every chapter published in this book has been scrutinized by our experts. Their significance has been extensively debated. The topics covered herein carry significant information for a comprehensive understanding. They may even be implemented as practical applications or may be referred to as a beginning point for further studies.

We would like to thank the editorial team for lending their expertise to make the book truly unique. They have played a crucial role in the development of this book. Without their invaluable contributions this book wouldn't have been possible. They have made vital efforts to compile up to date information on the varied aspects of this subject to make this book a valuable addition to the collection of many professionals and students.

This book was conceptualized with the vision of imparting up-to-date and integrated information in this field. To ensure the same, a matchless editorial board was set up. Every individual on the board went through rigorous rounds of assessment to prove their worth. After which they invested a large part of their time researching and compiling the most relevant data for our readers.

The editorial board has been involved in producing this book since its inception. They have spent rigorous hours researching and exploring the diverse topics which have resulted in the successful publishing of this book. They have passed on their knowledge of decades through this book. To expedite this challenging task, the publisher supported the team at every step. A small team of assistant editors was also appointed to further simplify the editing procedure and attain best results for the readers.

Apart from the editorial board, the designing team has also invested a significant amount of their time in understanding the subject and creating the most relevant covers. They scrutinized every image to scout for the most suitable representation of the subject and create an appropriate cover for the book.

The publishing team has been an ardent support to the editorial, designing and production team. Their endless efforts to recruit the best for this project, has resulted in the accomplishment of this book. They are a veteran in the field of academics and their pool of knowledge is as vast as their experience in printing. Their expertise and guidance has proved useful at every step. Their uncompromising quality standards have made this book an exceptional effort. Their encouragement from time to time has been an inspiration for everyone.

The publisher and the editorial board hope that this book will prove to be a valuable piece of knowledge for students, practitioners and scholars across the globe.

Index